THERMAL SPRAYING METHODS FOR REPAIRING AND MANUFACTURING

PLASMA, ARC & FLAME SPRAY TECHNOLOGY

edited by
Greg Easter

Wexford
2008

INTRODUCTION TO THERMAL SPRAYING TECHNOLOGY

Preface

This book is divided into two main sections. The first is a simplified overview of thermal spraying, followed by a more detailed account of thermal spraying methods. The first section is based on a manual used by the United States Navy in the repair and refurbishing of marine engine parts. This manual has been substantially altered from its original content with updated references and numerous varied corrections, but still the frame of mind is in the use of thermal spraying as a repair method. Other sources incorporated into this manual include the Metco manuals and information from several other manufacturers of equipment for thermal spraying.

The second section of this book is taken largely from a recent U.S. Army manual that teaches thermal spray technology at a more detailed level. It covers both original manufacturing of new parts as well as some aspects of repair. Both sections contain useful and hard-to-find information.

Finally there is a third section with excerpts of the practical aspects of shipboard thermal spray methods, contributed by the National Steel and Shipbuilding Company of California.

The reader who wishes only to obtain an overview of the thermal spray process can find it in Section I without having to wade through the detail of the second half. The reader who is already familiar with the basics, and who desires a more in-depth understanding, will find Section II a useful reference.

Table of Contents

SECTION I

THERMAL SPRAY MANUAL

The Society of Naval Architects and Marine Engineers

Ray Travis
Charles Ginther
Steve Vittori
Mel Herbstritt
Steve Herbstritt
Neil Armentrout
Ken Avery
Bill Williams
Brian Lawlor
Tom Marsh
S.C. Nelson

Edited, revised and reset by
Wexford Press
2008

The following document is based in part on information obtained through the U.S. Navy.

DISCLAIMER

These reports were prepared as an account of government-sponsored work. Neither the United States, nor the United States Navy, nor any person acting on behalf of the United States Navy (A) makes any warranty or representation, expressed or implied, with respect to the accuracy, completeness or usefulness of the information contained in this report/ manual, or that the use of any information, apparatus, method, or process disclosed in this report may not infringe privately owned rights; or (B) assumes any liabilities with respect to the use of or for damages resulting from the use of any information, apparatus, method, or process disclosed in the report. As used in the above, "Persons acting on behalf of the United States Navy" includes any employee, contractor, or subcontractor to the contractor of the United States Navy to the extent that such employee, contractor, or subcontractor to the contractor prepares, handles, or distributes, or provides access to any information pursuant to his employment or contract or subcontract to the contractor with the United States Navy. ANY IMPLIED WARRANTIES OF MERCHANTABILITY AND/OR FITNESS FOR PURPOSE ARE SPECIFICALLY DISCLAIMED.

Futhermore, the publisher (Wexford Press) has produced this purely for academic information purposes.No portion is intended to be used as a manual for actual practice of any of the methods discussed.

1 FUNDAMENTALS OF THERMAL SPRAYING

THERMAL SPRAY PROCESS

Thermal spraying is a group of processes in which freely divided metallic or nonmetallic surfacing materials are deposited in a molten or semimolten state onto a prepared substrate to form a deposited coating. The thermal spray process normally is a cost-effective method for achieving equipment repair or improving new or used equipment performance. The process is an excellent repair method and should be considered in lieu of other processes. Some of the potential problems of competing processes include the following:

1. Welding existing components may cause unacceptable distortion, cracking, or heat treatment problems. Precise identification of the base metal is required for welding. Some alloys are difficult or impossible to weld. Welding frequently lowers the fatigue life of the component.

2. Electroplating may be too slow or too limited in buildup capability, and it is environmentally hazardous.

BENEFITS

Advantages of using the thermal spray process for machinery components include the following:

1. BETTER THAN NEW - Thermal sprayed coating properties can be tailored to suit the application. Coatings (metallic and ceramic) can restore or attain desired dimensions, provide electrical or thermal shielding or conduction, or improve the resistance to abrasion, corrosion, or high temperatures. Thermal sprayed coatings can provide superior service to the extent that some coatings can extend the service life of mating components. For some of these applications, facilities are thermal spraying new components before being placed into service, to further extend equipment service life. In short, thermal sprayed coatings are capable of making many components significantly better than new.

2. NO DISTORTION OR HEAT AFFECTED ZONE-Although the substrate is normally preheated, the thermal spray process is still considered a metallurgically cold process. This means that, in most applications, no distortion or heat affected zone is created. **This is due to the fact that components generally are kept cooler than 400°F during the** entire spraying process.

3. VARIED SERVICE APPLICATIONS - Current thermal spiny applications in the shipbuilding and ship repair industry include thermal spraying of metallic materials on pump shaft bearing journals, turbine rotor shaft labyrinth seal areas, impeller fit areas, oil fit areas and various other types of applications. Ceramic coatings have been applied to the packing areas of valve stems and other types of packing and seal areas, giving those areas a smoother longer lasting surface than the substrate material thus enhancing the service life of that component. Thermal sprayed coatings can also serve for clearance control by applying abradable coatings at contact points.

4. **PART AVAILABILITY** - Parts availability plays a significant role in determining whether or not to use thermal spray repair methods. For some older machinery components, parts may simply be unavailable for a variety or reasons. For other items, ordering lead times may be unacceptable due to ship availability requirements. For still other high-cost items, new prices may be several times the cost of thermal spray repairs. In all of these cases, cost savings quickly add up, especially when factoring in the cost of ships unavailable for service commitments.

THERMAL SPRAY SYSTEMS

Shipbuilding and ship repair facilities use five primary thermal spray processes. These processes are the plasma powder process, the arc wire process, the flame powder process, the flame wire process, and the high velocity oxygen fuel process.

All thermal spray processes utilize five basic components. These components vary widely from one manufacturer or process to another, yet these components perform the same basic function (See Figure 1.1).

1. The Thermal Spray Gun is the heart of the system. This is where the heating zone forms and where the spray media enters the spray stream.

2. The Control System provides monitoring and control capabilities for the spray system. Operators use it to set and monitor gas flows and energy levels.

3. The Thermal Energy Supply may be electricity, or combustion from fuel gasses.

   Compressed gasses include compressed air for cooling and/or atomizing, fuel gasses, and inert gasses used for various functions, such as for plasma and powder carrier gas for powder feeders.

5. The final basic component is the Spray Media Feed System. This system delivers the spray media, usually in either wire or powder form, into the heating zone at a freely controlled rate. The controls for the feed system may be stand-alone or integrated with the spray gun control system.

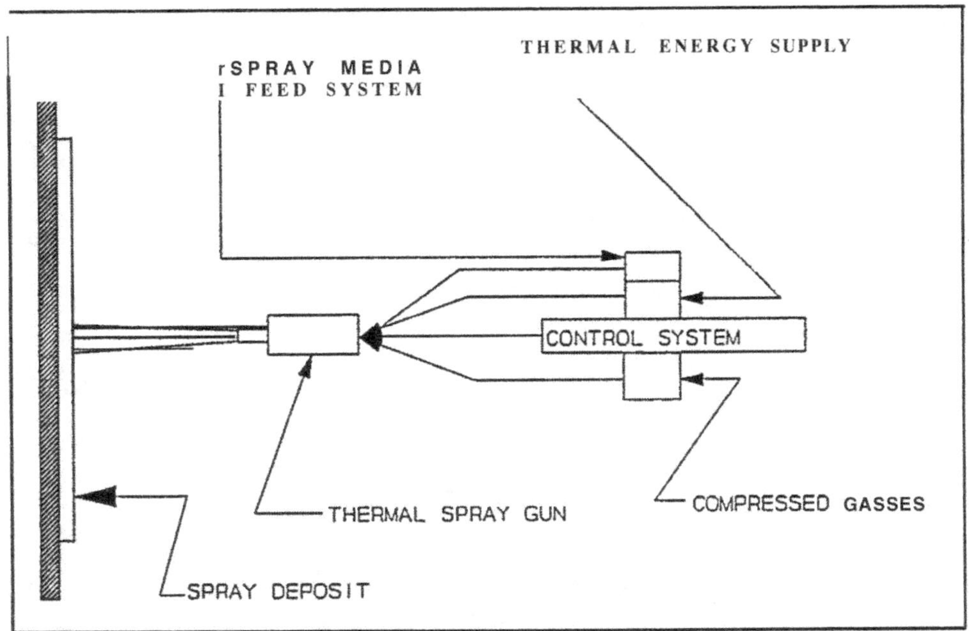

Figure 1.1 THERMAL SPRAY SYSTEM

PLASMA POWDER PROCESS

With the plasma powder process, an arc that is created between a tungsten electrode and a copper nozzle produces the heating zone. The tungsten electrode and nozzle are located in a conducting channel inside the plasma powder gun. A gas or gas mixture passes through this channel. The arc in the gas channel excites the gas into a plasma state, creating temperatures higher than can be obtained with any type of oxygen/fuel mixture. The gas or gas mixture then exits the gun forcing the plasma arc outside the gun. A powder is fed into the arc where it is melted and propelled at sonic or higher velocities to the substrate (See Figure 1.2).

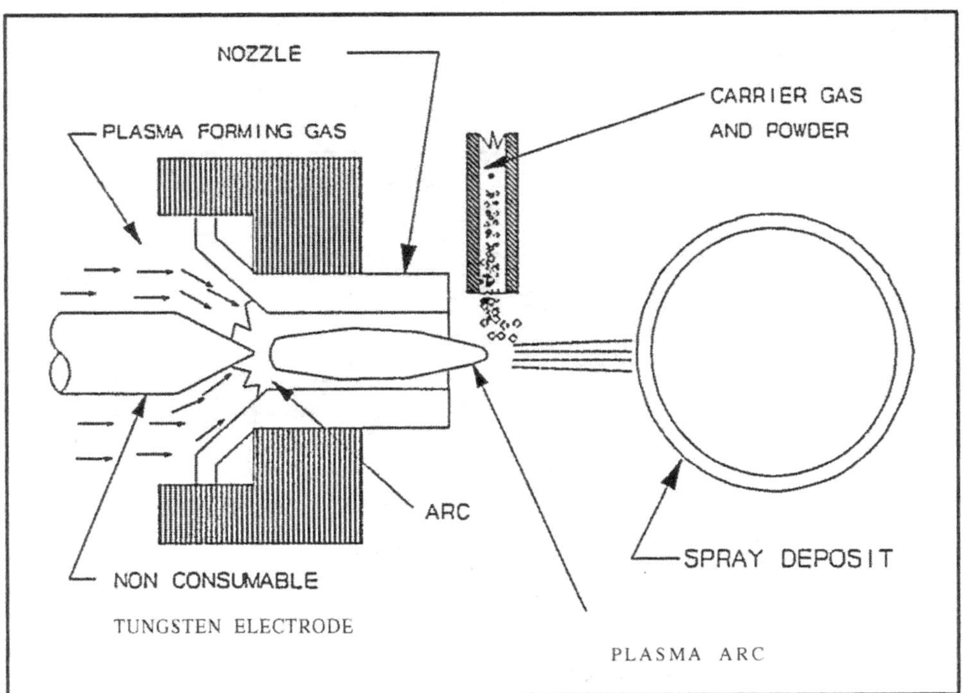

Figure 1.2 PLASMA POWDER PROCESS SCHEMATIC

ARC WIRE PROCESS

The heating zone with the arc wire process forms when an electric arc passes between two continuously fed metallic wires. Compressed air or an inert gas then atomizes and propels the molten material to the substrate (See Figure 1.3).

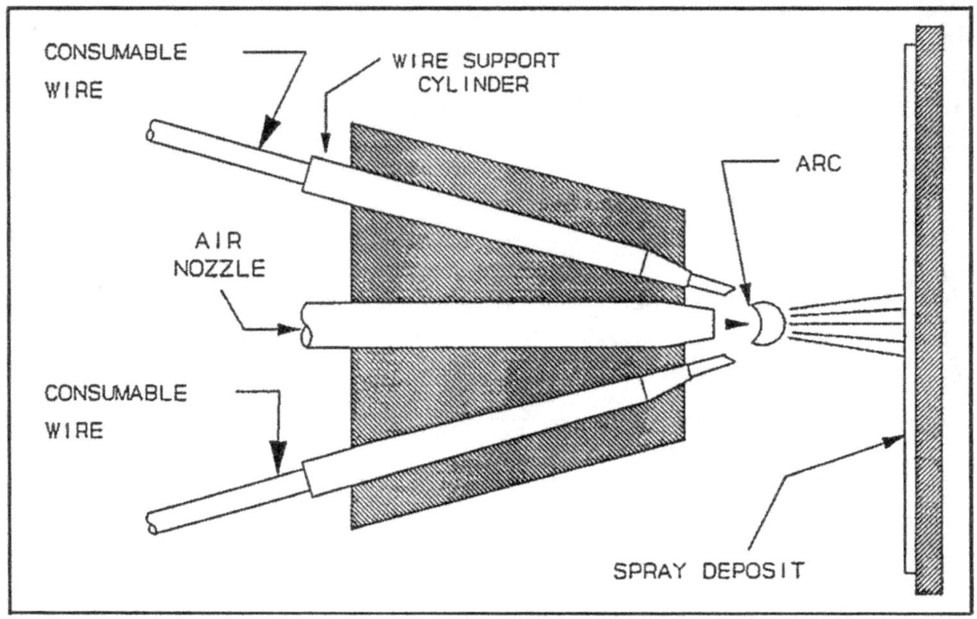

Figure 1.3 ARC WIRE PROCESS SCHEMATIC

FLAME WIRE PROCESS

The flame wire process uses an oxygen/fuel flame to create the heating zone. A wire continuously fed into the heating zone is melted, atomized, and then propelled onto the substrate by the force of the burning gasses and compressed air (See Figure 1.4).

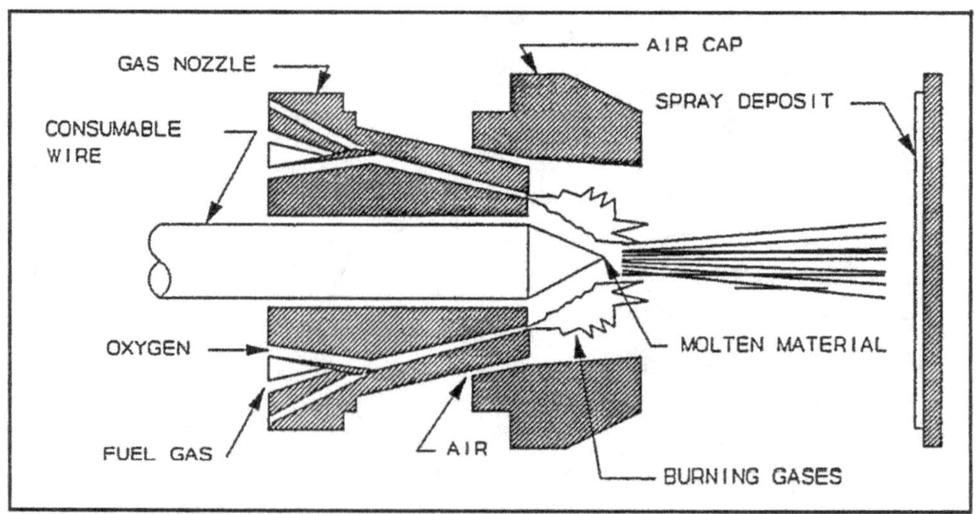

Figure 1.4 FLAME WIRE PROCESS SCHEMATIC

FLAME POWDER PROCESS

The flame powder process uses an oxygen/fuel flame to create the heating zone. Powder enters the heating zone and melts. The molten material is then propelled to the substrate by the force of the burning gasses and compressed air (See Figure 1.5).

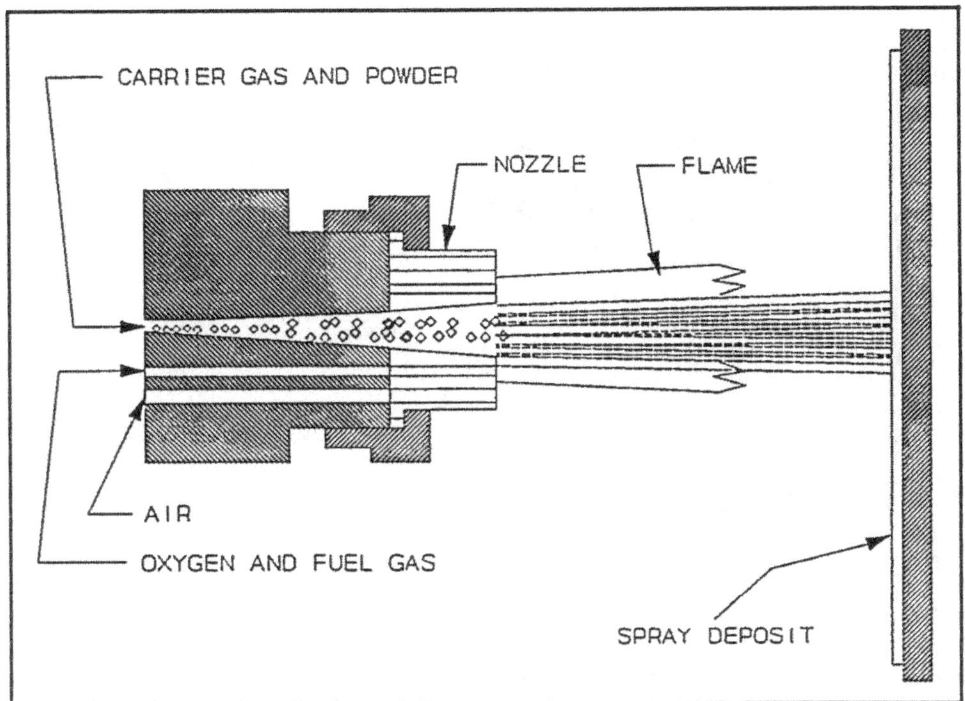

Figure 1.5 FLAME POWDER PROCESS SCHEMATIC

HIGH VELOCITY OXYGEN FUEL (HVOF) PROCESS

The HVOF process is one of the most recently developed of the thermal spray processes. This process uses the combustion of oxygen and fuel (typically propane, propylene, or hydrogen) mixtures at high pressures. This mixture of oxygen and fuel gas bums and is accelerated to supersonic speeds through special shaped nozzles. Thermal spray powder is injected into this constricted flame. The powder then accelerates in the high velocity flame (4500-7000 feet per second) (See Figure 1.6). When the thermal spray material collides with the substrate it plasticizes, flattens and adheres to form a coating. Some of the advantages of this process are due to the fact that it uses relatively low thermal energy coupled with high kinetic energy. This process produces coatings with low porosity, low oxide content, low residual stress, and exceptionally high bond strengths. The HVOF process typically produces coatings with reduced changes in the metallurgical phase composition of the thermal spray material when compared to the other common shipyard thermal spray processes.

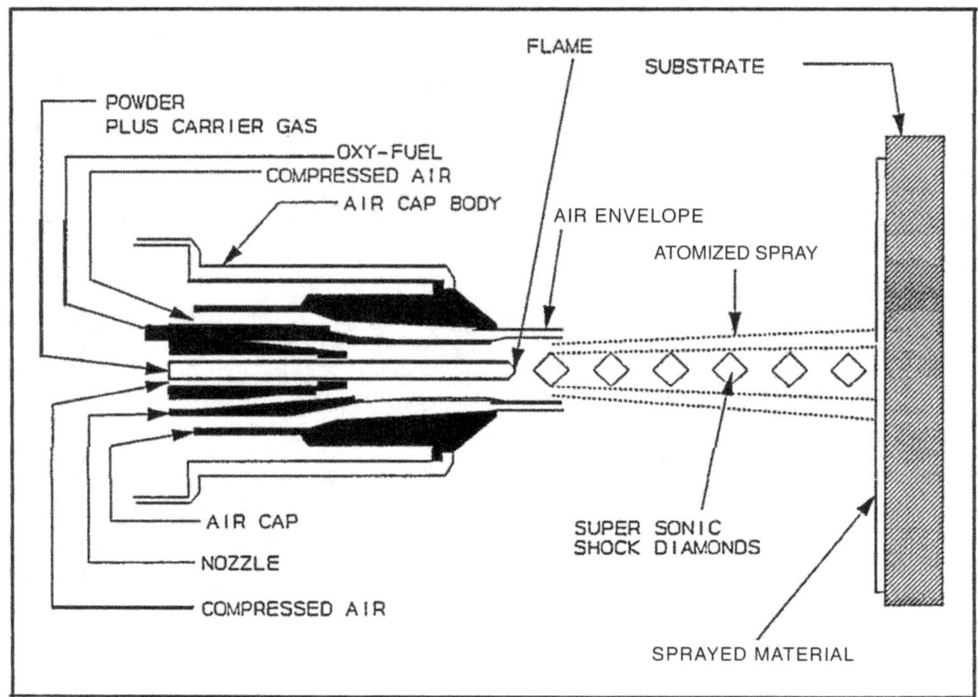

Figure 1.6 HVOF PROCESS SCHEMATIC

OTHER METHODS

Various industries apply thermal spray materials and methods that are not utilized by shipyards or marine repair facilities, which was the source of much of the material in this first half of the book. New processes and equipment that will provide superior coatings for future applications are constantly being developed by manufacturers of thermal spray equipment, as well as those companies that offer thermal spraying as a service. So again the reader is reminded that this first section is intended only to impart an understanding of the overall process.

LIMITATIONS

As with any other repair method, Thermal Sprayed Coatings do have some limitations. Thermal Sprayed Coatings have a significantly lower tensile strength than typical structural materials. Therefore, if undercutting for surface preparation reduces component strength to unacceptable levels, alternative repair or preparation methods must be used. Thermal Sprayed Coatings do not typically stand up to point or line loading, such as would be produced in a ball or roller bearing race. Peening is another example of this type of loading. Any application which seems to fall into this category should either be reconsidered or subjected to laboratory or in-service testing.

CONCLUSION

The Thermal Spiny Process provides many benefits for new and used components. These benefits are capable of producing significant cost savings for ship repair facilities, especially over the long term. Any facilities desiring information or assistance with thermal spray technology can contact the authors of this manual at Puget Sound Naval Shipyard.

2 COMPARISON OF METHODS & EQUIPMENT

INTRODUCTION

The principle types of thermal spray processes being used in industry are the plasma powder, arc wire, flame wire, high-velocity oxygen fuel, and flame powder processes. There are many modifications and innovations (e.g. twin wire arc, high-velocity spray, etc.) but the underlying principles are the same as the ones shown here.

PROCESS SELECTION

When choosing a thermal spray process for a specific application, numerous factors which affect quality and productivity must be compared. This can become a complex decision due to the number of conflicting advantages and limitations which each process may possess for a given situation. Each process can be ranked in terms of its deposition rate in pounds of thermal spray material deposited per hour. However, there are other factors which must be considered. For instance, a new process may have a high deposition rate, but may require extensive training and research and development for procedure qualifications. With some processes, the cost to set up in a field situation may be prohibitive. Process selection should be based on personnel skill and their knowledge of available processes. Many factors must be taken into account for the best process selection.

EQUIPMENT AVAILABLE

Metco (now Sulzer Metco) has been manufacturing thermal spray equipment for more than 70 years. A guide to their most current equipment is available free online at:

http://www.sulzermetco.com/en/Portaldata/13/Resources/2_products_services/2.1_thermal_spray/EquipmentGuide_Web_Prelim.pdf

TABLE 4.1

THERMAL SPRAY PROCESS	PLASMA POWDER	ARC WIRE	FLAME POWDER	FLAME WIRE	HVOF
QUALITY	EXCELLENT	EXCELLENT	GOOD	GOOD	EXCELLENT
SPRAY RATE	GOOD	EXCELLENT	MEDIUM	MEDIUM	GOOD
PORTABILITY	POOR	EXCELLENT	EXCELLENT	GOOD	EXCELLENT
EQUIPMENT MAINTENANCE	MEDIUM	LOW	LOW	MEDIUM	HIGH
OPERATOR SKILL REQUIRED	HIGH	LOW	LOW	MEDIUM	MEDIUM
COST	HIGH	LOW	LOW	LOW	HIGH

Most large thermal spray shops have access to the equipment required for all of the processes discussed here. TABLE 4.1 is a comparison guide for the most routinely used processes. The following pages summarize these common thermal spray processes.

PLASMA POWDER

With the plasma powder process, an arc that is created between a tungsten electrode and a copper nozzle produces the heating zone. The tungsten electrode and nozzle are located in a conducting channel inside the plasma powder gun. A gas or gas mixture passes through this channel. The arc in the gas channel excites the gas into a plasma state, creating temperatures higher than can be obtained with any type of oxygen fuel mixture. The gas or gas mixture then exits the gun forcing the plasma flame outside the gun. A powder is fed into the flame where it is melted and propelled at sonic or higher velocities to the substrate.

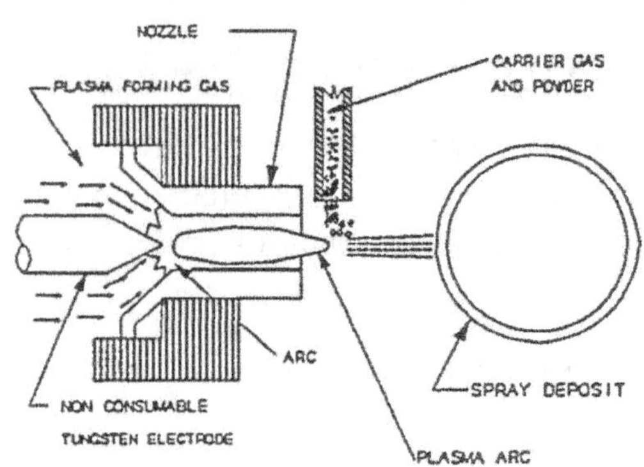

Figure 4.1

ADVANTAGES	LIMITATIONS
Reliable	Not as portable as some of the other thermal spray processes
Capable of spraying all thermal spray powders, including ceramics	High operating cost
Produces coatings with high bond strengths, low porosity, and low oxides	High start-up cost
Easily adapted to robotics	Requires highly skilled thermal spray operators
Low substrate heat input	High intensity ultra-violet light can quickly cause flash bums to inadequately protected skin or eyes
Good deposition rates	May require water cooling
	High noise levels generally require double hearing protection for operators and any other personnel who must work nearby

ARC WIRE

The heating zone with the arc wire process forms when an electric arc passes between two continuously fed metallic wires. Compressed air or an inert gas then atomizes and propels the molten material to the substrate.

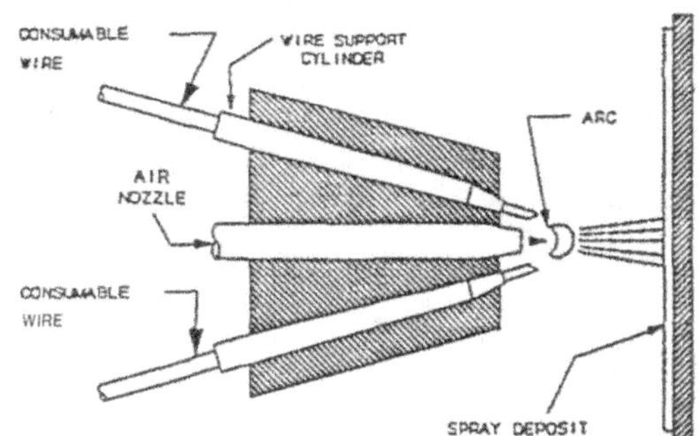

Figure 4.2

ADVANTAGES	LIMITATIONS
Very portable	Can not thermal spray ceramics (except for a few specialty cermet wires designed for producing non-skid coatings)
Easy to operate	Higher porosity and oxides than plasma powder and high velocity oxygen fuel coatings
High deposition rates	High intensity ultra-violet light can quickly cause flash bums to inadequately protected skin or eyes
Relatively inexpensive to purchase and operate	
Produces coatings with high bond strengths	
Has the ability to use low cost materials such has welding wires	
Very low heat input	

FLAME WIRE

The flame wire process uses an oxygen fuel flame to create the heating zone. A wire continuously fed into the heating zone is melted, atomized, and then propelled onto the substrate by the force of the burning gasses and compressed air.

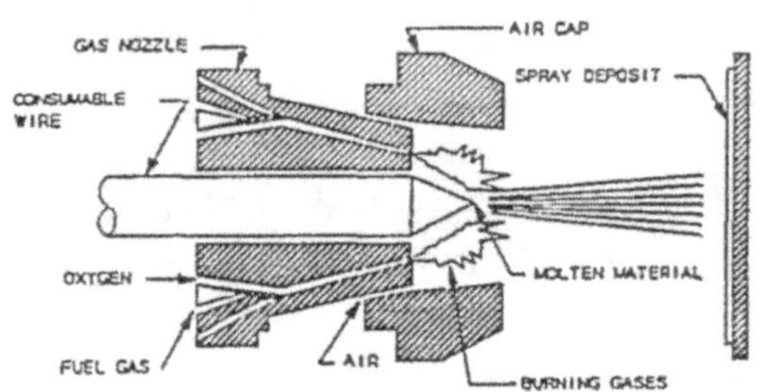

Figure 4.3

ADVANTAGES

Medium deposition rates

Capable of spraying most thermal Spiny wires

Relatively inexpensive to purchase and operate

Very portable

Easy to operate

LIMITATIONS

Medium ratio of heat conveyed into the substrate

Unable to spray ceramic materials without some special equipment

Low velocity spiny stream

Lower bond strengths than arc wire, plasma powder, and high velocity oxygen fuel

FLAME POWDER

The flame powder process uses an oxygen fuel flame to create the heating zone. Powder enters the heating zone and melts. The molten material is then propelled to the substrate by the force of the burning gasses and compressed air

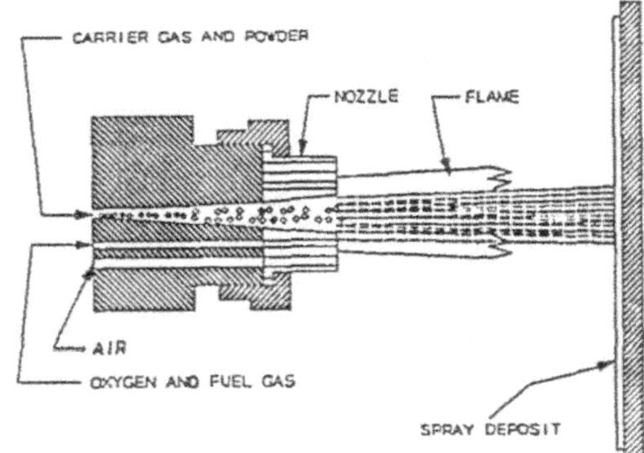

Figure 4.4

ADVANTAGES	LIMITATIONS
Relatively high deposition rates	High ratio of heat conveyed into the substrate
Capable of spraying most thermal spray powders	Unable to spray some ceramic materials
Low noise levels	Low velocity spray stream
Relatively inexpensive	
Very portable	
Easy to operate	

HIGH VELOCITY OXYGEN FUEL

The HVOF process is one of the most recently developed of the thermal spray processes. This process uses the combustion of oxygen and fuel (typically propane, propylene, or hydrogen) mixtures at high pressures. This mixture of oxygen and fuel gas bums and is accelerated to supersonic speeds. Thermal spray powder is injected into this constricted flame. The powder then accelerates in the high velocity flame (4500 - 7000 feet per second) When the thermal spray material collides with the substrate it plasticizes, flattens and adheres to form a coating.

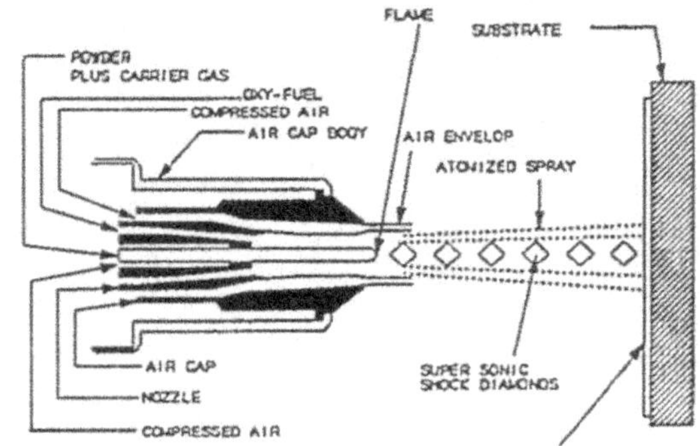

Figure 4.5

ADVANTAGES	LIMITATIONS
Low thermal energy	High noise levels generally require double hearing protection for operators and any other personnel who must work nearby
High kinetic energy	Relative expensive
Coatings with low porosity, low oxide content, low residual stress, and exceptionally high bond strengths	Requires highly skilled operators
Reduced changes in the phase composition of the thermal spray material	

3 SELECTION OF COATING MATERIALS

COATING SYSTEMS

Many factors must be considered when selecting a coating material. In addition to the basic type of coating material chosen (e.g. nickel alloy, alumina, etc.) there are a vast array of proprietary materials that are known only by their performance specifications and either a trade name or, more frequently, a number designated by the manufacturer.

COATING SYSTEM SELECTION

Because of the plethora of coating materials on the market, with new materials being introduced every year, there is no way to summarize the pros and cons of every material for every possible application. The single most popular process is plasma spraying, and the most popular type of material is ceramic. The following are the traditional ceramic coating material types:

- Aluminum Oxide - Resistant to abrasion, friction and oxidation up to 1100°C. Used mostly in coating machine parts for the textile industry.
- Chrome Oxide - Corrosion resistant material used extensively in the chemical industry for pump parts and seals. Resists oxidation up to 1000°C.
- Zirconium Oxide - Because of its wear resistance and heat resistance, this is extensively used in coating jet turbine parts as a TBC (thermal barrier coating.) Resists oxidation to 900°C.
- Titanium Oxide - Also very resistant to both abrasion and oxidation. Many applications.

COATING SELECTION CHARTS

There are many suppliers of thermal spray coating materials. The following websites will give you some idea of the variety and applications of such materials, and typical numbering systems:

http://www.sermatechmaterialsdivision.com/resourceCenter/technical_datasheets.html

http://www.precisioncoat.com/coatServHome.htm

If you are still confused about which thermal spray process and/or coating material is best for your application, you can consult experts online at no cost, as well as read what others have asked, at the following website:

http://www.gordonengland.co.uk/sef/thermal-spray-misc-questions-t-232.html

The following pages are a series of charts created by the Naval Surface Warfare Center in Annapolis, and Puget Sound Shipyard. The extensive experience of those Naval Facilities in the thermal spray field has led to these choices for maritime machinery component applications. Your applications will probably vary considerably, but at least this will provide some idea of the factors taken into consideration when selecting thermal spray materials.

TABLE 5.1 CARBON, LOW ALLOY, STAINLESS STEEL, AND CAST IRON						
COATING SYSTEM	STEAM		FRESH/SALT WATER		AIR/OIL	
	FIT	* SEAL	FIT	* SEAL	FIT	* SEAL
PP-25	N	N	N	N	N	N
444	3	N	3	N	1	N
445	N	N	N	N	N	N
447	4	N	4	N	2	N
21021	2	N	2	N	4	N
21031	1	N	1	N	3	N
PP-25/130	N	N	N	N	N	N
444/130	N	N	N	3	N	1
445/130	N	N	N	N	N	N
447/130	N	N	N	4	N	3
21021/130	N	N	N	2	N	4
21031/130	N	N	N	1	N	2
PP-25/143	N	N	N	N	N	N
444/143	N	3	N	7	N	5
445/143	N	N	N	N	N	N
447/143	N	4	N	8	N	7
21021/143	N	2	N	6	N	8
21031/143	N	1	N	5	N	6

1-8 = NUMBER LISTED IN COLUMN - RANKING BASED ON MIL-STD-1687 OR PSNS QUALIFIED PROCEDURE AND EXPERIENCE - THE NUMBER 1 INDICATES THE BEST COATING SYSTEM SELECTION FOR THE LISTED APPLICATION. THE COATING SYSTEM WITH THE LARGEST NUMBER IN THE COLUMNS COULD BE USED, BUT WOULD BE THE LAST RECOMMENDED SYSTEM FOR THE APPLICATION.

= POOR SELECTION BASED ON EXPERIENCE OR NOT PERMITTED

(*) = DOES NOT INCLUDE LABYRINTH SEALS

NOTE: WHEN EVER POSSIBLE GRINDING IS THE PREFERRED FINISHING METHOD FOR ALL THERMAL SPRAYED COATINGS.

TABLE 5.2
NICKEL BASE ALLOYS

COATING SYSTEM	STEAM		FRESH/SALT WATER		AIR/OIL	
	FIT	* SEAL	FIT	* SEAL	FIT	* SEAL
PP-25	N	N	1	N	1	N
444	N	N	N	N	N	N
445	N	N	N	N	N	N
447	N	N	3	N	2	N
21021	N	N	2	N	3	N
21031	N	N	N	N	N	N
PP-25/130	N	N	N	1	N	1
444/130	N	N	N	N	N	N
445/130	N	N	N	N	N	N
447/130	N	N	N	3	N	2
21021/130	N	N	N	2	N	3
21031/130	N	N	N	N	N	N
PP-25/143	N	N	N	4	N	4
444/143	N	N	N	N	N	N
445/143	N	N	N	N	N	N
447/143	N	N	N	6	N	5
21021/143	N	N	N	5	N	6
21031/143	N	N	N	N	N	N

1-8 = NUMBER LISTED IN COLUMN - RANKING BASED ON MIL-STD-1687 OR PSNS QUALIFIED PROCEDURE AND EXPERIENCE - THE NUMBER 1 INDICATES THE BEST COATING SYSTEM SELECTION FOR THE LISTED APPLICATION. THE COATING SYSTEM WITH THE LARGEST NUMBER IN THE COLUMNS COULD BE USED, BUT WOULD BE THE LAST RECOMMENDED SYSTEM FOR THE APPLICATION

= POOR SELECTION BASED ON EXPERIENCE OR NOT PERMITTED

(*) = DOES NOT INCLUDE LABYRINTH SEALS

NOTE: WHEN EVER POSSIBLE GRINDING IS THE PREFERRED FINISHING METHOD FOR ALL THERMAL SPRAYED COATINGS.

TABLE 5.3 COPPER BASE ALLOYS						
COATING SYSTEM	STEAM		FRESH/SALT WATER		AIR/OIL	
	FIT	* SEAL	FIT	* SEAL	FIT	* SEAL
PP-25	N	N	N	N	N	N
444	N	N	N	N	N	N
445	N	N	1	N	1	N
447	N	N	N	N	N	N
21021	N	N	N	N	N	N
21031	N	N	N	N	N	N
PP-25/130	N	N	N	N	N	N
444/130	N	N	N	N	N	N
445/130	N	N	N	1	N	1
47/130	N	N	N	N	N	N
21021/130	N	N	N	N	N	N
21031/130	N	N	N	N	N	N
PP-25/143	N	N	N	N	N	N
444/143	N	N	N	N	N	N
445/143	N	N	N	2	N	2
447/143	N	N	N	N	N	N
21021/143	N	N	N	N	N	N
21031/143	N	N	N	N	N	N

1-8 = NUMBER LISTED IN COLUMN - RANKING BASED ON MIL-STD-1687 OR PSNS QUALIFIED PROCEDURE AND EXPERIENCE - THE NUMBER 1 INDICATES THE BEST COATING SYSTEM SELECTION FOR THE LISTED APPLICATION. THE COATING SYSTEM WITH THE LARGEST NUMBER IN THE COLUMNS COULD BE USED, BUT WOULD BE THE LAST RECOMMENDED SYSTEM FOR THE APPLICATION

= POOR SELECTION BASED ON EXPERIENCE OR NOT PERMITTED

(*) = DOES NOT INCLUDE LABYRINTH SEALS

NOTE: WHEN EVER POSSIBLE GRINDING IS THE PREFERRED FINISHING METHOD FOR ALL THERMAL SPRAYED COATINGS.

TABLE 5.4
WIRE COATINGS FOR CARBON, LOW ALLOY, STAINLESS STEEL, NICKEL BASE ALLOYS, AND COPPER BASE ALLOYS

ARC-WIRE FLAME-WIRE COATING SYSTEM	WEAR RESISTANT	CORROSION RESISTANT	SINGLE POINT TOOLING RECOMMEND	SUBSTRATE MATERIALS
410	Yes	No	No	CARBON, LOW ALLOY, STAINLESS STEEL
309	Yes	No	No	CARBON, LOW ALLOY, STAINLESS STEEL
METCOLOY #2	Yes	No	No	CARBON, LOW ALLOY, STAINLESS STEEL
70S-4	Yes	No	No	CARBON, LOW ALLOY STEEL
EN-60	No	No	Yes	NICKEL BASE ALLOYS
C-22	Yes	Yes	No	STAINLESS STEEL & NICKEL BASE ALLOYS
625	Yes	Yes	No	STAINLESS STEEL & NICKEL BASE ALLOYS
HAYNES-75-B	Yes	No	Yes	CARBON, LOW ALLOY STEEL
AMPCO-10	No	Yes	Yes	COPPER BASE ALLOYS

4 PREPARATION OF SURFACES FOR THERMAL SPRAYING

INTRODUCTION

A clean surface is a requirement for all types of thermal spraying. In the case of repairing old parts, extensive cleaning of the surface may be required. Methods fall into two categories, solvent cleaning, and abrasive blast cleaning.

SOLVENT CLEANING

Remember that water is a solvent. Water with a detergent added is usually the first step in cleaning a surface to be thermal sprayed. This removes both dirt and oil. Steam cleaning may be required, depending on the nature of the surface and the extent of contamination.

Hydrocarbon petroleum-based solvent is almost always employed as a final step before spraying. All traces of the solvent are baked off, typically at 600°C to ensure that no trace of the solvent or moisture remains trapped on the surface. If spraying will not immediately follow, then the surface must next be protected against any further contamination with a clean dry wrapping. Even the amount of oil that transfers from a single fingerprint will interfere with the integrity of the sprayed coating.

ABRASIVE BLAST CLEANING

Alumina is the most frequently used material for this process. Several manufacturers produce trade name alumina materials specifically for cleaning parts to be flame sprayed. One popular and excellent choice is Metco's METCOLITE abrasive.

ADDITIONAL DETAILS

Further details about cleaning parts for thermal spraying are provided in the second half of this book in Section 6.

5 THERMAL SPRAY OPERATIONS

INTRODUCTION

Producing a quality thermal sprayed coating requires a high degree of control over each stage of operation. These stages include application and coating selections, cleaning, undercutting, masking, surface preparation, thermal spraying, and finishing. Application of a quality coating requires trained and skilled mechanics who know exactly what must be accomplished during each of these stages, particularly during the thermal spray operation. This section provides information about setting up and operating thermal spray and related equipment during the actual spray operation. This Section was developed to help trainees, journeyman level mechanics, and other interested personnel improve their skills and gain insight into the day-to-day operation of a thermal spiny facility.

EQUIPMENT SETUP

The frost step of the spray operation is usually performed before anchor tooth blasting begins. Tools must be staged, and rotating equipment must be adjusted. In some cases, components must be temporarily mounted in the futures to assure correct fit, alignment and clearances. Finally, the spray equipment must be started up. There are many equipment manufacturers who sell thermal spray systems. It is beyond the scope of this manual to list the steps for setting up all of the systems available on the market. Set up procedures are provided by the equipment manufacturers in their operating manuals and training programs. Trained and qualified operators must know how to set up and use the spray equipment used at their facilities. Appendix (7.1) is an example of a start up procedure that can be used for one model of a Plasma Spray System. When using other equipment or processes, the equipment manufacturer's recommended start up procedures should be followed.

TOOLS

Thermal spray operators use a variety of tools in the performance of their duties. An assortment of generic and specialized hand tools, such as hammers, screwdrivers, wrenches, and cutting tools accumulates for the various purposes of building, adjusting, and maintaining all of the fixtures, masks, and spray equipment. In addition, four special quality assurance tools are required for controlling and assuring a quality thermal sprayed coating.

1. Compressible plastic film and the related thickness gage are used to measure the anchor tooth height and to give an indication of the anchor-tooth blast quality.
2. Vernier or dial caliper or micrometers are used to measure diameter changes. Highly accurate readings are essential for controlling deposition rates (with ceramics, as refine as 0.0005" per pass may be required). This can be a real challenge on rough surfaces which may expand and contract throughout the spray process.
3. Temperature monitoring instruments (usually contact pyrometers) help operators control the temperature of a component.

As operators gain experience, both in general and with specific components, the number of stops for taking temperature readings may be reduced; however, frequent readings are still necessary to keep positive control of component and coating temperatures.

4. A 10X magnifier is used as an aid during visual inspections to analyze coating quality. Specific details of visual inspections are covered in the Quality Assurance section of this manual.

TURNING FIXTURE

Any machinery component should be, if at all possible, sprayed using automatic or semi-automatic equipment with a gun fixture mounting. This is to maintain the close control of parameters so necessary for the types and amounts of coating buildup typical for these applications. Relative movement between the gun and substrate must be accurately controlled. The angle of the thermal spray gun should be as close as possible to 90° to the substrate. The gun angle must never be less than 45° to the substrate. In general, turning fixtures used for thermal spraying can be anything that will hold and rotate the component, have a mount to hold the spray gun, meet the speed and feed requirements of the procedure, and will not contaminate the area to be thermal sprayed. Several manufacturers sell turning fixtures designed specifically for thermal spray operations. Many thermal spray facilities use a lathe for their turning fixture. These lathes are usually equipped with a self centering head chuck and a live, spring loaded center in the tail stock. If possible, the undercutting, spraying, and finishing should be accomplished on the same centers of the component. An ideal fixture is a CNC Machining Center designed to undercut/prep/spray/finish. Although few facilities can achieve that goal, most can approach it by undercutting and finishing on the same centers or using only one setup and providing centers in the component for the spray facility to use in this operation. To determine the rotating speed for the turning fixture, the following formula can be used:

Diameter for area to be thermal sprayed = D (in inches)

Surface feet per minute = SFPM

Revolutions per minute $= \frac{12 \times SFPM}{\pi \times D}$

The traverse speed range for the turning fIXture is set to the parameters listed in the procedure.

When final surface preparation is completed, the component must be inspected to insure correct anchor tooth profile and that 100% of the area to be sprayed is 100% grit blasted. For more information see Preparing a Machinery Component for Thermal Spraying, Section 6, of this Manual.

Before placing the machinery component in the turning fixture, the condition of the masking must be examined, and if required (keyway masks, for example), changed to insure that no damage will be caused during or after the spraying process. When placing the component into the turning fixture, care must be taken so that the area to be sprayed is not contaminated. If handling requires touching the area to be sprayed after final surface preparation, operators must wear clean white cotton gloves. Even the normal amount of oil from human skin may cause a non-bonded area between the substrate and the bond coat.

If the component must be moved to another building or through a contaminated environment, the prepared surface should be covered to protect it from oxidation and airborne contamination. Also, to help minimize oxidation, the thermal spraying must betin within two hours after completion of the final surface preparation.

PREHEATING

After the component is placed in the turning fixture, measuring for the calculation for the expansion of the substrate material can be started. These calculations must be made because most thermal spray coatings have thickness limitations. With ceramic coatings and some metallic coatings, these thickness limitations are critical to the success of the spray application. When applying these types of coatings, the expansion of the machinery component must be taken into consideration. To calculate the expansion of the component and the diameters necessary to achieve the thickness required for the bond coat and the final coat, the formulas of TABLE 7.1 can be used. Note that these calculations are based on undercut limits which are based on the limitations of coating thickness. The final measurements for buildup calculation purposes are taken just before the first pass is sprayed.

If a component is to be sprayed without preheat, the expansion allowance must be estimated. For most single coating applications, 0.001" for every inch of component diameter (rounded to the nearest inch) provides a good ballpark number. If a more accurate value is needed, expansion tables are available which can be used to create a temperature correction curve. Contact local engineers for this information.

BUILDUP CALCULATIONS SHEET
TABLE 7.1

SYMBOLS	
C=GRIT BLASTED DIA. (AMBIENT) H=GRIT BLASTED DIA (225-250°F)	E=H-C
E=EXPANSION ALLOWANCE B=BOND COAT GOAL DIA.	B=F+E-O.020
S=STOP SPRAY GOAL DIA. F=FINISH DIA.	S=F+E+O.020

Machinery components must never be sprayed if the substrate temperature is below 60°F or if the substrate temperature is less than 10°F above the dew point of the ambient air. The normal preheat temperature range for machinery components is 200°F to 250°F. Preheating is accomplished to remove moisture from the surface to be sprayed and may also minimize coating stress due to expansion of the component. When the plasma powder process is used, the plasma spray system can be utilized to preheat the component. If one of the oxygen fuel spray systems is used, the gas flame can be utilized for preheat. However, if preheating is done with a gas flame, the flames must never be directly applied onto the area to be sprayed (preheating the area to be sprayed directly with an oxy/fuel flame promotes oxidation of the prepared surface).

When using the arc wire spray process, most procedures permit a 60° F preheat; when preheat is required for the arc wire process, either the plasma torch or an oxygen fuel torch may be employed for preheating. Temperature sticks, or any other device that would contaminate the area to be thermal sprayed, must never be used to check the temperature of the component. A contact pyrometer is the typical instrument used by thermal spray facilities to check preheat and interpass temperature.

THERMAL SPRAYING

After preheating, the masking should be inspected again and replaced if necessary. Any masking repairs must be performed in such a way as to prevent contamination of the prepared surface. See Section 6, Preparing A Machinery Component For Thermal Spraying for further details. Masking should be monitored all during the spraying process and repaired or replaced when required. The spraying operation should only be interrupted to check temperature, measure thickness, repair or replace masking, change spraying material and parameters from the bond coat to the final coat, and to permit cooling to prevent overheating. (During spraying, the temperature of the substrate should not exceed 400°F or the tempering/aging temperature, whichever is lower.) The thermal spray operator must have sufficient training and experience to prevent excessive heat input to the machinery component. Accelerated cooling, such as a blast of clean air, carbon dioxide or other suitable gas, can be used if it is not applied directly upon the area being sprayed. Liquids must never be used for cooling the component. Once spraying starts, coating thickness per pass and the ratio of traverse rate to rpm become critical parameters. Surface speed is critical only as a starting point. If the surface feet per minute and traverse speed are initially set too slow, over-heating of the coating may take place. Applying the thermal spray material too thick per pass causes internal stresses that can lead to cracking and delamination between the layers of the coating. If the feed rate is set too fast in relation to the surface feet per minute, an uneven coating may be applied (See Figure 7.1).

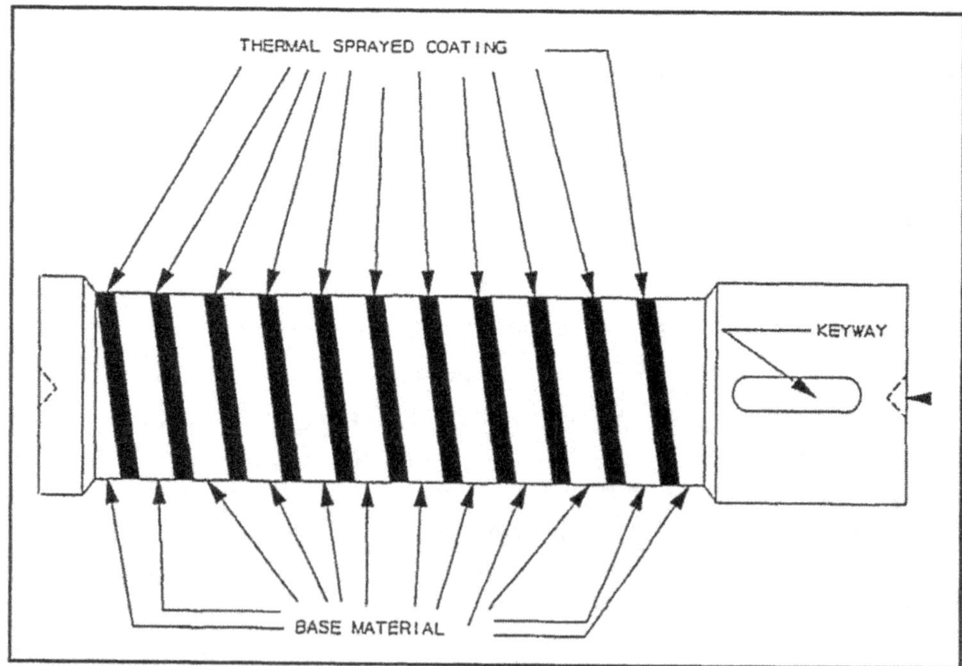

Figure 7.1 Barber Pole Effect

The thermal sprayed coating should by applied in multiple passes. Thicknesses of passes must be in accordance with the procedure for the coating system being applied. The entire area being sprayed should be coated before proceeding to the next pass. Measurements of the coating layers are required to insure the coating thickness per pass and speed and feed parameters that are listed in the procedure are being met.

If carbon inserts are used, the part must be carefully monitored to check to see if the coating is peeling from the insert. If this happens, stop and break off the coating flakes to prevent excessive shadowing of the adjacent area.

If trouble arises during the coating process, spraying must stop, and the problem must be freed. If the problem has caused a defect in the coating, the coating must be removed, and the surface blasted and thermal sprayed again.

At all times during spraying, the parameters must be monitored to insure that the procedure is being followed accurately.

COOLING & SEALING

When thermal spraying is completed, the machinery component is allowed to cool. During the cooling cycle, the machinery component should, if practical, be left rotating slowly in the turning fixture to help prevent warpage. If it is necessary to cool the component quickly, clean compressed cooling air can be used. If compressed cooling air is used, component rotation should be maintained. Compressed cooling air must be maneuvered to obtain a uniform cooling rate over the entire section. Liquids must not be used for cooling. If the component must be removed from the turning fixture, support it in a vertical position so that heat can escape evenly around the circumference of the spray zone. When the coating temperature cools to 125° F, a sealer is applied. Seal coatings must be applied in a well ventilated area. If the component is being sealed in the turning fixture, the same ventilation system used for spraying can be used during the sealing process. Phenolic resins in solution, such as Metco AP sealer, are generally used by most ship repair and ship manufacturing facilities. Sealers can be applied by using any of the common painting methods, as appropriate for the surface to be sealed.

DEMASKING

In a sense, demasking can be said to begin as soon as blasting is completed. In general, surfaces used for mounting the component are masked only for anchor tooth blasting. The rest of the masking is frequently left in place during preheating in order to help retain heat. Once spraying starts, significant portions of the masking are generally removed to aid in drawing heat away from the spray zone, thus reducing the time, required for interpass cool down. After spraying is completed, all or portions of the masking may be removed to aid cool down and/or to ease the application of sealer. The decision of when and how much masking to remove is a judgement call aided by experience. The following considerations will provide help in making these decisions.

1. A fringe benefit of masking is the physical protection provided while moving components from one place to another.
2. Removal of masking from mounting points is usually necessary after blasting in order to accurately mount the component for spraying.
3. The heat retention capability of blast masking, although limited, can sometimes significantly improve time requirements for preheat. This is especially true when the masking traps air. Component configuration and spray zone location play a significant role in this function.
4. Once preheat temperature is reached, the spray process continues to add heat to the component. Removal of some blast masking can, in some cases, significantly speed up cool down between passes as well as after spraying is complete.
5. Any procedure which uses cooling air during spraying (this includes all arc spray procedures) tends to have little or no need for auxiliary cooling.
6. Any areas which have the potential of receiving direct impact of spray material must be left masked until spraying is completed.
7. At times, application of sealer is considerably easier if shadow masks are moved away from the spray zone or removed completely.
8. Masking tape left in place during cool down generally makes for an easier job of applying sealer.
9. Tape left in place during sealing is usually easier to remove before the sealer dries.

In general, masking is removed in the opposite order of application and with little need for consideration of damage to the coating (other than common sense caution). However, a few precautions, are in order.

1. Thermal spray coatings do not holdup well to impacts; never strike the coating with a hammer or other hard tool.
2. Thermal sprayed coatings do not hold up well to point loading; never use a prybar to force a masking fixture to move.
3. If spray material should lap over masking tape, do not attempt to pull the tape directly away from the component. The sprayed coating can crack or chip back into the undercut zone, ruining the coating. Instead, chip the overlapped coating away from the spray zone as shown in Figure 7.2.
4. If spray material should lap over masking fixtures, the overlapping coating must be ground down using a hand held grinder until the fixture is freed. Do not force the grinder. Do not permit the grinder to heat spots more than warm to the touch. If at all possible, direct grinding force away from the spray zone and toward the masking fixture. Never force the fixture to move.

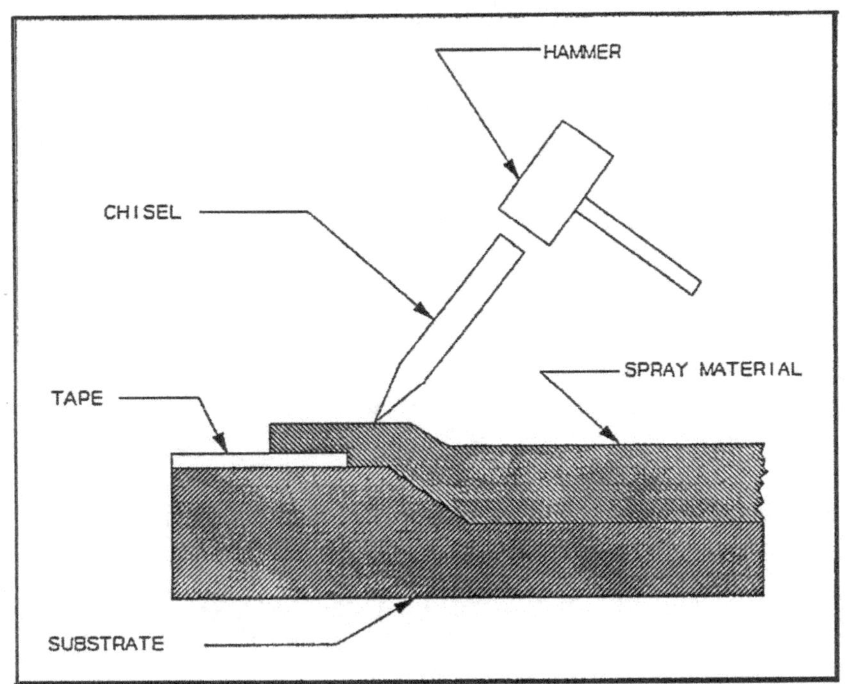

Figure 7.2 METHOD OF TAPE REMOVAL

APPENDIX 7.1
A TYPICAL PLASMA POWDER SYSTEM START UP PROCEDURE *

1. Connect and turn on gas, observing local safety requirements.
2. Install nozzle, powder port, and air jets on gun, as required by the procedure.
3. Mount thermal spray gun in the traversing gun mount, pointing it into the spray booth.
4. On the control panel, place the automatic/manual selector switch in manual.
5. On the control panel, place the Argon/Nitrogen selector switch to the gas being used.
6. Press power on button.
7. Check flows and pressures.
   Press and hold purge button.
   a Adjust primary flow to specified level, reading flow valve at the center of the ball float.
   c. Set the secondary flow to 27-29.
   d. Check primary and secondary pressures at 7M control panel.
   e. If pressure of either gas is different than specified, release purge button, adjust pressure at the gas bottle, and repeat step 7.
   f. When pressures are correct, close secondary flow valve and maintain purge flow for 5 seconds.
   g. Release purge button.
8. Check gun interior.
   a. Press test button and hold it
   b. After 5 seconds, inspect interior of the spray gun using a mirror (and wearing safety glasses).
   c. Observe presence of high frequency arc between electrode and nozzle.
   d. Observe that interior of spray gun and face of mirror are free of any sign of moisture.
   e. Observe satisfactory condition of nozzle and electrode.
   f. Release test button.
   g. If the gun fails step 8.c, 8.d, or 8.e, correct the condition and repeat step 8.
9. Set up powder feeder.
   a. Press auto-run/manual-standby button.
   b. Adjust carrier gas flow to specified level.
   c. Check powder unit for leaks.
   d. Set spray rate control dial to specified value specified for the procedure to be sprayed.
   e. Press auto-run/manual-standby off button.
10. Adjust secondary flow needle valves.
    a. Open secondary flow valve 2 turns.
    b. Press auto-run/manual-standby button.
    c. Observe flow rate sequence.
    d. If partial flow is 15 and full flow is 29-30, skip to step 10.i.
    e. Close full flow needle valve.
    f. Adjust partial flow needle valve for flow rate of 15.
    g. Adjust full flow needle valve for flow rate of 29-30.

APPENDIX 7.1
A TYPICAL PLASMA POWDER SYSTEM START UP PROCEDURE *

h. Close secondary flow valve.
i. Press auto-run/manual-standby off button.

11. Light off spray gun and set parameters.
    a. Turn arc current control to 3.
    b. Place automatic/manual selector switch in automatic.

<WARNING>

DO NOT PERMIT SECONDARY GAS TO
FLOW WHEN THE THERMAL SPRAY GUN IS OPERATING BELOW 300 AMPS
(AFTER GUN IS LIT)

    c. Press auto-run/manual-standby button--gun should light.
    d. Dial in desired parameters manually.

NOTE

If amps and gas flows set to specified levels produce out-of-specification voltage, shut down system and replace nozzle and/or electrode. Then press power on and repeat steps 4,7,8, and 11.

    e. Press auto-run/manual-standby off button, and wait for system to cycle off.

12. Test for repeatability and smoothness of operation by lighting off spray gun and shutting down. Fine tune settings, if necessary.
13. Load hoppers(s) with desired powder.
    Thoroughly mix powder.
    a. Add powder to hopper.
    c. Test feed rate using an approved method (See Enclosure (7.l)).
14. Thermal spiny bend test and record results.

* This start up procedure is for a typical Metco 7M Plasma Powder System. Thermal spray facilities should always use their equipment manufacturer's start up procedure.

POWDER FEED RATE CONVERSION CHART

Factor (O.132) X Grams Per Minute = Lbs per Hour

Grams/Min	#/HR	Grams/Min	#/HR	Grams/Min	#/HR
1	.132	41	5.41	81	10.69
2	.264	42	5.50	82	10.82
3	.396	43	5.68	83	10.96
4	.529	44	5.81	84	11.09
5	.661	45	5.94	85	11.22
6	.792	46	6.06	86	11.35
7	.925	47	6.20	87	11.48
8	1.056	48	6.34	88	11.62
9	1.189	49	6.47	89	11.75
10	1.320	50	6.60	90	11.88
11	1.45	51	6.73	91	12.01
12	1.58	52	6.86	92	12.14
13	1.71	53	7.00	93	12.28
14	1.85	54	7.13	94	12.41
15	1.98	55	7.26	95	12.54
16	2.15	56	7.36	96	12.67
17	2.24	57	7.51	97	12.80
18	2.38	58	7.65	98	12.94
19	2.51	59	7.79	99	13.06
20	2.64	60	7.91	100	13.20
21	2.77	61	8.05	101	13.33
22	2.91	62	8.19	102	13.46
23	3.04	63	8.30	103	13.60
24	3.17	64	8.45	104	13.73
25	3.30	65	8.56	105	13.86
26	3.43	66	8.70	106	14.00
27	3.56	67	8.85	107	14.12
28	3.70	68	8.96	108	14.26
29	3.83	69	9.10	109	14.39
30	3.96	70	9.24	110	14.52
31	4.10	71	9.37	111	14.65
32	4.25	72	9.50	112	14.78
33	4.35	73	9.64	113	14.92
34	4.49	74	9.75	114	15.05
35	4.62	75	9.90	115	15.18
36	4.75	76	10.03	116	15.31
37	4.87	77	10.16	117	15.44
38	5.01	78	10.30	118	15.58
39	5.15	79	10.43	119	15.71
40	5.27	80	10.56	120	15.84

40 SECOND METHOD

1. Weigh empty container or zero the scale with empty container on the platform.
2. Collect powder in container for 40 seconds using the carrier flow rate listed in the procedure.
3. Weigh container with powder.
4. Subtract empty weight from weight with powder (Note If scale was zeroed with container on platform, this step can be skipped. This is powder weight.
5. Double weight of powder.
6. Move decimal one place to the left. This is reading in pounds per hour.

60-SECOND METHOD

1. Weigh empty container.
2. Collect powder in container for 60 seconds using the carrier flow rate listed in the procedure.
3. Weigh container with powder.
4. Subtract empty weight from weight with powder. This is powder weight.
5. Multiply by .132 or use chart to determine #/HR.

EXAMPLE OF 40 SECOND METHOD:

A. Empty container weighs 15.0 grams.
B. Container with powder weigh 65 grams.
c. Container & powder = 65
Empty weight - 15
Result 50
(50 is the weight of the powder.)
D. Double weight of powder. 100
E. Move decimal point one
place to the left. = 10.0
(10 #/HR is the spray rate.)

EXAMPLE OF 60 SECOND METHOD:

A. Empty container weighs 15.0 grams.
B. Container with powder weigh 91 grams.
c. Container & powder = 91
Empty weight - 15
Result 76
(76 is the weight of the powder.)
.132 X 76 = 10.032
LBS/HR = 10.032

NOTE: On the chart, 76 Grams per Minute corresponds with 10.03 #/HR.

FINISHING OF THERMAL SPRAYED COATINGS

INTRODUCTION

To a great extent, the effectiveness of thermal sprayed coatings is dependent upon the finishing techniques employed. The fact that the coatings are not a homogeneous mass, but rather many particles bonded together, dictates that sprayed coatings be finished with wheels and techniques or cutting tools and parameters not normally used on similar material in wrought or cast form. Also, the interposed oxides cause tool wear and grinding problems common with harder materials. This causes difficulty for machinist and machine operators who are inexperienced with the finishing of coatings. Improper machining methods can cause particle pull out, cracking, disbanding at the substrate bond coat interface, delamination between coating layers, and/or a rough finish. By carefully observing the rules governing wheel or cutting tool selection, and by employing proper grinding techniques or machining parameters, the finishing of thermal spray coatings can become relatively trouble free.

METHODS

Most thermal sprayed marine machinery components are finished by either using carbide cutting tools, high speed cutting tools, or by grinding. Wet grinding is the preferred and most common finishing method for most thermal sprayed coatings. This is especially true with hard, wear resistance coating like ceramics and carbides. When grinding these harder coatings, it is essential that the correct grinding wheel and grinding techniques are selected.

GRINDING

The American Welding Society professes that there are four general rules in grinding wheel selections. There are as follows:

1. Dress the wheel as frequently as necessary in order to use the sharpest wheel possible. Sharp wheels cut rapidly without overheating that can cause de-lamination.
2. Choose wheels with structures and grades that provide free cutting action.
3. Choose the grinding wheel bond type best suited to the operation and equipment.
4. Know the equipment - both machines and wheels,

Paramneters for grinding include the following;

1. Wheel Speed
2. Work Speed
3. Area of Contact
4. Wet Grinding
5. Wheel Dressing

RECOMMENDED GRINDING TECHNIQUES:

1. Seal the coating surface before starting the grinding process. Most shipbuilding and ship repair facilities use a phenolic sealer, such as Metco AP, for sealing thermal sprayed coatings.
2. Use softer, free cutting wheels. Chances of burnishing and particle pull-out will be greatly reduced.
3. Maintain the wheel face in a clean and sharp condition. Dress the wheel frequently to keep the face free-cutting. Grinding wheels with dull abrasive will create friction and heat.
4. Use coarse grit wheels for maximum stock removal and fine grit wheels for finishing. Attempting to generate fine finishes with coarse grit wheels that have been dressed closed can result in particle pull-out, smearing, and burnishing.
5. Use light cuts. Sprayed coatings are usually very thin. Excessive grinding pressure can cause delamination of the sprayed surface or particle pull-out.
6. Do not spark out on the final pass; this tends to glaze or dull the wheel face.
7. Grind wet whenever possible. Improved finishes, less chance of burnishing, less heat-checking, and less contamination will result.
8. Use freer grit wheels on dense, hard-to-penetrate sprayed coatings.
9. Use narrower wheels on machines with low horse-power and for more rapid stock removal of hard materials.
10. Always keep the coating under compression. By cutting down through the sprayed surface towards the substrate, delamination and particle pull-out will be minimized.
11. On encountering problems with a given wheel, experiment with wheel speeds, feed rates, work speeds, and-dressing techniques. Changes in variables can have a significant effect on stock removal rates and finishes.

Grinding wheel selection is based on the size of the component, required surface finish, hardness and structure of the coating, and the condition and capabilities of the grinding machine. Table 8.1 list grinding wheel recommendations for some of the thermal sprayed materials that are used for repairing machinery components.

TABLE 8.1
GRINDING WHEEL RECOMMENDATIONS

MATERIAL	WHEEL SPECIFICATIONS	GRINDING APPLICATION			
		CENTERLESS	CYLINDRICAL	INTERNAL	SURFACE
Alumina Chromium Oxide Titania Alumina Zirconia Titania, Yttria	Abrasive type:	C (Green)	C (Green)	C (Green)	C (Green)
	Grit size:	(180 μm)	80 (180 μm)	(180 μm)	(180 μm)
	Grade	G	G	J	F
	Bond:	Vitrified	Vitrified	Vitrified	Vitrified
	Diamond type:	Mfd Ni Clad	Mfd Ni Clad	Mfd Ni Clad	Mfd Ni Clad
	Grit size:	(120 μm)	(120 μm)	150/180 (100/83 μm)	125 (120 μm)
	Grade:	R	R	FUN	R
	Concentration	75	75	100	75
	Bond:	Reainoid	Reainoid	Reainoid	Resinoid
Chromium Cobalt Nickel	Abrasive type:	CIA	CIA	CIA	CIA
	Grit size:	(250 μm)	(250 μm)	80 (180 μm)	46 (340 μm)
	Grade:	J	J	L	H
	Bend:	vitrified	vitrified	vitrified	Vitrified
Molybdenum	Abrasive type:	C (Black)	C (Black)	C (Black)	C (Black)
	Grit size:	(250 μm)	(250 μm)	80 (180 μm)	(180 μm)
	Grade:	I	I	N	H
	Bond:	vitrified	vitrified	vitrified	Vitrified
Stainless steel (400 series)	Abrasive type:	A	A	A	A
	Grit size:	(250 μm)	(250 μm)	80 (180 μm)	(340 μm)
	Grade:	J	J	L	H
	Bond:	vitrified	Vitified	Vitified	Vitified
High nickel, alloys and Stainless steel (300 Series)	Abrasive type:	CIA	CIA	CIA	CIA
	Grit size:	(250 μm)	(250 μm)	80 (180 μm)	(340 μm)
	Grade:	J	J	J/L	H
	Bond:	Vitrified	Vitrified	vitrified	Vitrified

TOOLING

Carbide tools have been found to be very satisfactory for machinable thermal sprayed coatings. Carbides cutting tools are required for the harder materials (See Table 8.2 for carbide tools machining recommendations). High speed steel cutting tools can be used on some the softer coatings, such as brass and bronze. (See Table 8.3 for high speed single point steel tools machining recommendations.) Just as in grinding, proper tool selection and cutting parameters are critical to the final quality of the coating. See Figures 8.1 and 8.2 for tool angles for carbide tools and the recommended grind angles for high speed tool bits.

Machining of a thermal sprayed coating should start with the removal of the overspray area at the end of the coated area (See Figure 8.3.).

The rules for machining thermal sprayed coating are at follows;

1. Use a properly sharpened cutting tool.
2. Use correct speeds and feeds.
3. Use correct tool angles.
4. Use correct tool configuration.
5. Do not hog-off the coating.

NOTE After final machining has been completed, any keyway or holes that were in the thermal spray should be beveled back at the edges. This can be accomplished by using a file with a material that is harder than the thermal sprayed coating.

CONCLUSION

To assist the machinists in the finishing of thermal sprayed coatings, it is recommended that each thermal spray facility purchase a copy of the American Welding Society's "Thermal Spraying - Practice, Theory, and Application". This publication can be ordered by calling (800) 443-9353, Ext 280. Machining parameters can also be found in the Flame Spray Handbook published by the Sulzer Metco Company.

Table 8.2
Carbide Tools Machining Recommendations

DEPOSITED THERMAL SPRAYED COATING	speed (SFPM)		(IPR)	
	Rough	Finish	Rough	Finish
Iron Base				
C .35; P .02; S .02; Mn .5; Cr 13.00; Si .5; Fe bal.	30-40	30-40	.004	.003
Mn .5; C .10; Fe bal. Mn .6; C .23; Fe bal.	75-100 50-75	75-100 50-75	.006 .004	.003 .003
C .15; Mn 8.5; Ni 5.10 Cr 18.0; Mn .7; Fe bal.	100-125	125-175	.006	.003
C .80; Mn .7; Fe bal. C .04; Mn 2.0; Ni 4.0; Cr 1.5; Mo 1.5; Fe bal.	30-40	30-40	.004	.003
Copper Base				
Al 9.5; Fe 1.0; Cu bal.	250-300	300-350	.006	.003
Nickel Base				
Ni 95; Al 5.	200-250	250-300	.004	.002

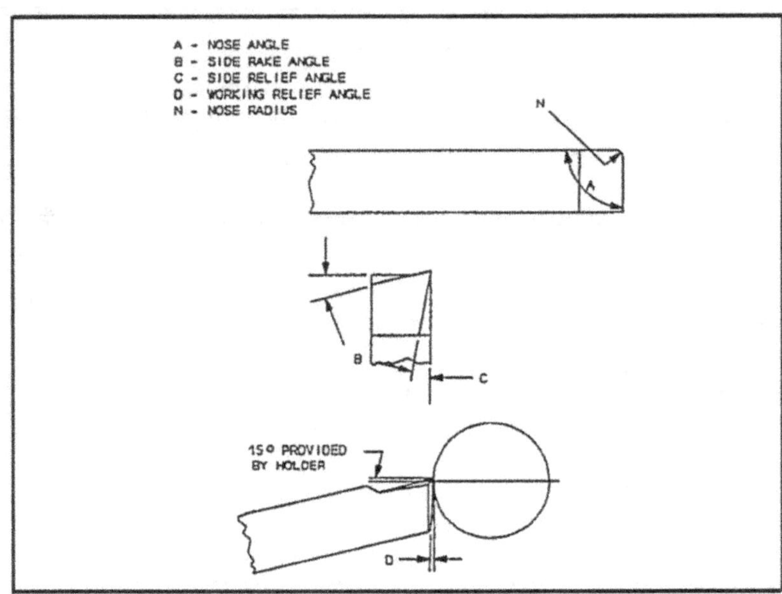

Figure 8.1 HIGH SPEED TOOL BITS GROUND AT ANGLES RECOMMENDED FOR MACHINING OF THERMAL SPRAYED MATERIAL

TOOL No. 1	A=80° B=0° C=10° N=0.030 in (0.76 mm)	D	
TOOL No. 2	A=80° B=10° C=10° N=0.030 in (0.76 mm)	D	AS LITTLE
TOOL No. 3	A=80° B=15° C=10° N=0.040 in (1.00 mun)	D	AS POSSIBLE

ALL TOOLS GROUND FOR USE IN ARMSTRONG TYPE HOLDER

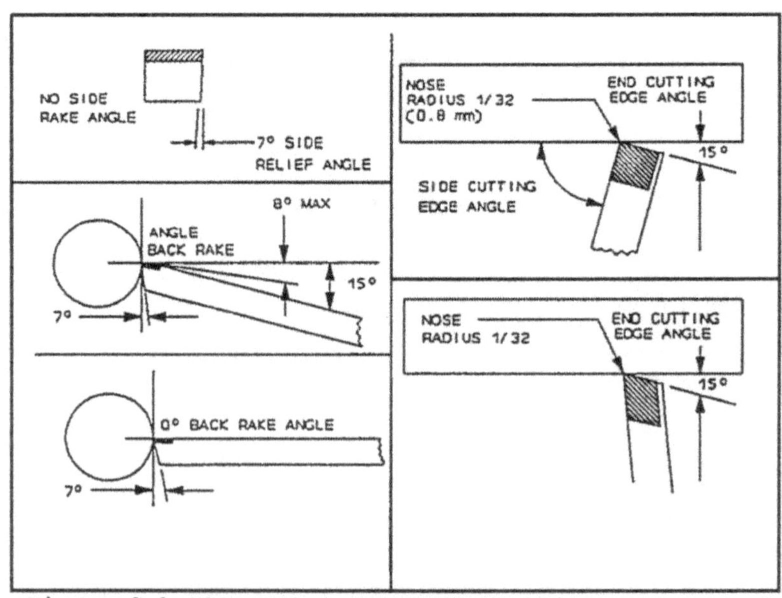

Figure 8.2 TOOL ANGLE FOR CARBIDE TOOLS

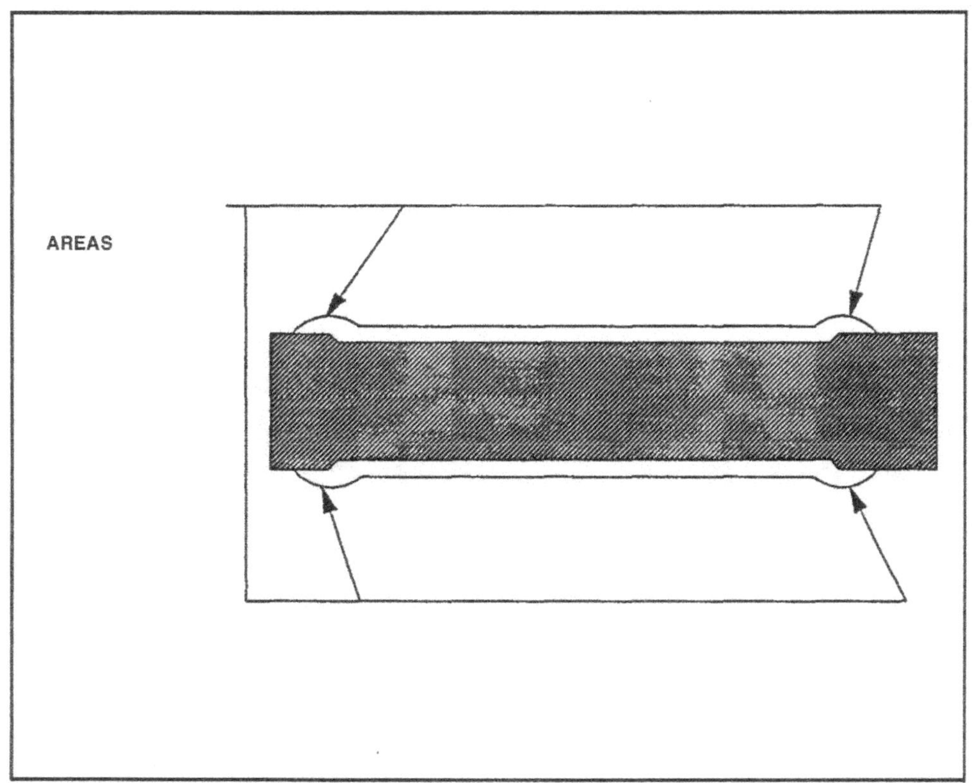

Figure 8.3 OVERSPRAY AREAS

7 PREPARING MACHINE PARTS FOR REPAIR BY THERMAL SPRAYING

INTRODUCTION

This section contains the approved methods used by Naval Facilities for the repair of marine equipment. Preparation of components for thermal spray repair involves four operations:

1. Cleaning, which is both the first operation and also repeated between subsequent operations.
2. Undercutting, in which all of the worn metal is cut away. For cylindrical parts, this is usually performed on a lathe.
3. Masking, which is done to protect sensitive parts of the component from grit blast damage and to insure that spray material covers only the intended areas.
4. Abrasive Blasting, which roughens the surface just before thermal spraying. Note that this is also called Anchor Tooth Blasting by some technicians.

CLEANING METHODS

Cleanliness of the area to be thermal sprayed has a direct influence on how well the coating adheres to the substrate material. All oil, grease, rust, scale, paint, or any other substance that would inhibit the thermal spray coating from adhering to the substrate must be removed prior to final surface preparation. The machinery component should be cleaned, both before and after undercutting. Also, it must be cleaned again just before final surface preparation. The methods currently used to clean machinery components in the maritime industry include baking, solvent washing, vapor decreasing, mechanical cleaning, and abrasive blasting.

1. Solvent washing is the most commonly used method for cleaning maritime machinery components that are to be thermal sprayed.
2. Vapor decreasing is used by some facilities as their main method to remove contaminants from components that are to be thermal sprayed.

< WARNING AND PRECAUTIONS>

Due to the flammable and toxic nature of most solvents, proper precautions must be followed during solvent cleaning. Precautions shall also be taken to protect any parts which may be attacked by the solvents. Do not clean previously thermal sprayed parts with acids or other corrosive fluids, as deterioration of the existing coating may result. If corrosive cleaners must be used, pre-existing thermal sprayed surfaces shall be completely protected or resprayed. Follow safety precautions listed on material labels and/or material safety data sheets.

3. Heat cleaning may be used on porous materials that have been contaminated with grease or oil. It is best to solvent clean before heat cleaning starts. Castings should be baked for four hours to char and/or drive out the foreign material from the pores. Steel castings should be heated at 650° F maximum. Aluminum castings should be heated to no more than 300° F; age hardened alloys must be kept below the aging temperature.

If there is any possibility that heat cleaning may change the existing heat treatment on the component or that distortion may occur, a welding engineering review should be considered before heat cleaning.

4. Abrasive blasting (some times called corrosion removal blasting) may be used to remove heavy or insoluble deposits. Inexpensive, non-reusable blasting grit is recommended for this purpose. Another good source of grit is partially broken down aluminum oxide which has been recycled from anchor-tooth blasting operations. For delicate surfaces, non-abrasive blast media may work better.
5. Replacements for chlorinated solvents have been developed by a number of companies. These solvents are designed to be significantly less hazardous to personnel and to the environment than the chlorinated solvents.
6. Solvent cleaning just prior to masking is important in order to permit tape or masking compounds to adhere as well as possible to areas which must be protected from blast and spray streams.

UNDERCUTTING

Undercutting on maritime machinery components that are to be thermal sprayed is accomplished to remove existing coatings, to remove damaged or contaminated base material, and to allow for a uniform finished coating thickness. Because most machinery components thermal sprayed at marine repair facilities are cylindrical in shape, a lathe is normally used for undercutting. To help insure uniform coating thickness, undercutting and finishing processes should be performed on the same centers. The undercut design is usually completed by the thermal spray operator or thermal spray planner when filling out Section II of the Thermal Spray Job Control Record (See Enclosure 2.2 in Section 2) (Normally, engineers and machinists have not been ttained in undercut design requirements). A typical undercut design is shown in Figure 6.1.

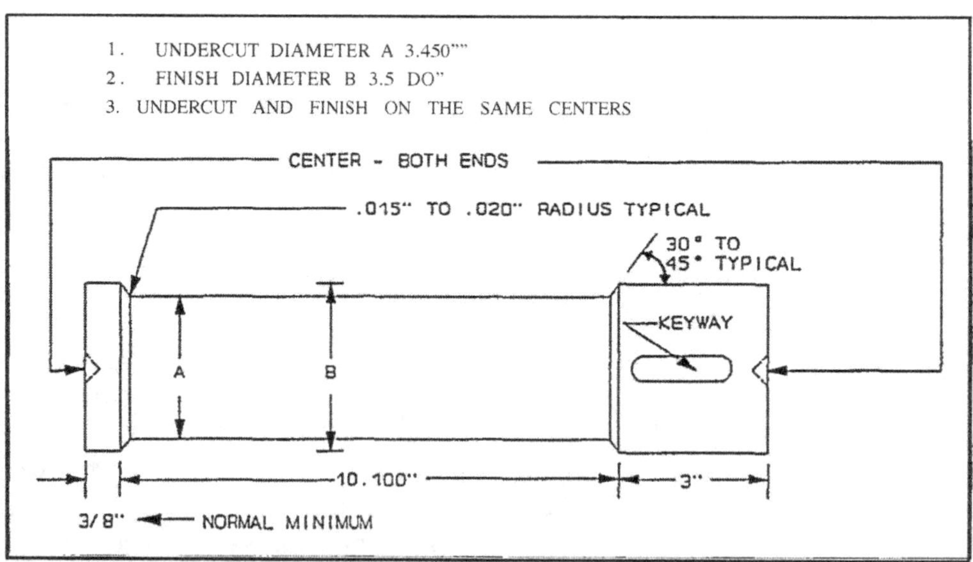

Figure 6.1 TYPICAL UNDERCUT DESIGN

NORMAL UNDERCUT GUIDELINES (FOR SHAFTS AND OUTSIDE DIAMETERS)

1. A radius of 0.015" to 0.020" should be cut at the comers of the undercut.

2. The shoulders should typically be no closer than 3/8" from an unprotected outside corner. (See page 6.7 for special conditions, and for exceptions to this general guideline.)
3. At each end of the undercut section on the machinery component, shoulders shall be cut at an angle of 30 to 45 degrees measured from the axis of the component (See Figure 6.2).

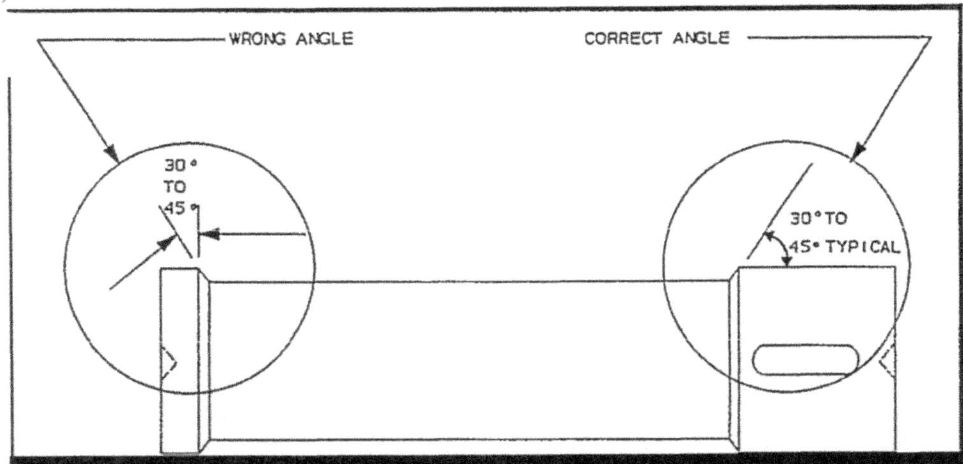

Figure 6.2 ANGLE OF SHOULDER

4. The depth of the undercut determines the thickness of the finished machined thermal sprayed coating. The optimum thickness of the sprayed coating is determined by the type of component, by the coating function, and by application area duty and the contraction rate of the chosen material(s). The component type and service conditions are the major factors in determining deposit thickness. These conditions can be reduced to the following recommendations:

- ☐ Recommended maximum depth of undercut for shafts is given in TABLE 6.1 in this section.
- ☐ If TABLE 6.1 permits more than .020" depth of undercut, a coating thickness limitation may apply. Check manufacturer's recommendation or TABLE 6.2 (Excerpted from MIL-STD-1687, TABLE 1) for normal coating thickness limits.
- ☐ If the sprayed deposit is not a wearing surface, no minimum thickness need be specified other than that the deposit should be of uniform thickness and meet the minimum thickness requirements of the applicable spray procedure.
- ☐ If the deposit is a wear surface, the minimum deposit thickness is determined by the maximum amount of wear the surface is allowed in service, plus .005" for each coating material to maintain adequate bonding and to distribute localized load forces. (Example, a valve stem: 0.005" Bond Coat, 0.005" Ceramic, plus 0.005" Ceramic Wear Allowance - Total, 0.015" minimum depth of undercut).

NOTE: If wear or undercut depth reduces the strength of the component enough to make it unusable for service, then thermal spray must be rejected as a repair option (See TABLE 6.1).

CONDITIONS TO AVOID

1. Dovetails or even angles steeper than 45° significantly reduce coating quality in these areas, and (in the case of dovetails) may even leave voids in areas shadowed from direct particle impact. These defects are likely to cause premature coating failure, and they may be exposed during finish machining operations. Figures 6.3 and 6.4 are examples of poorly designed undercuts.
2. Although keyways must sometimes be included within the spray area, they should be avoided, if possible, because of the stress on the keyway edge.
3. Taper and run out (eccentricity) must be strictly limited in the undercut area in order to minimize loss of shaft strength and to maintain even coating thickness. This is especially important when spraying coatings such as ceramics which have severely limited buildup tolerance.
    a. Maximum taper is typically 0.002".
    b. Maximum runout is typically 0.001".

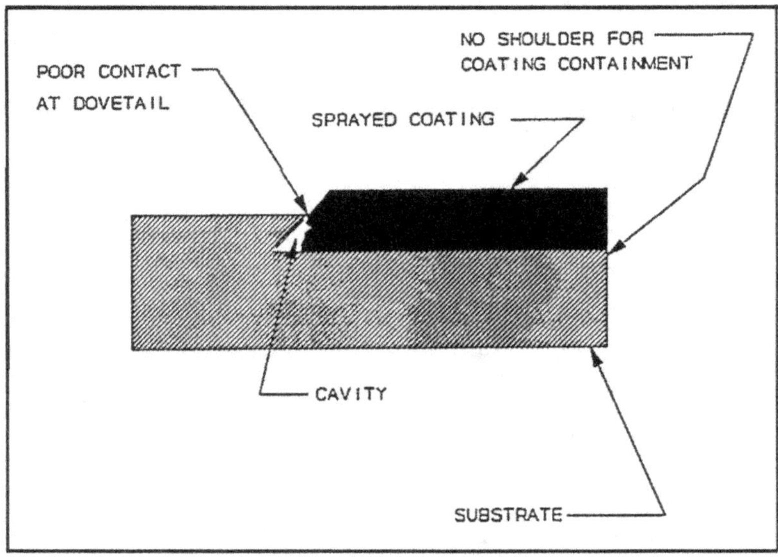

Figure 6.3 POORLY DESIGNED UNDERCUT

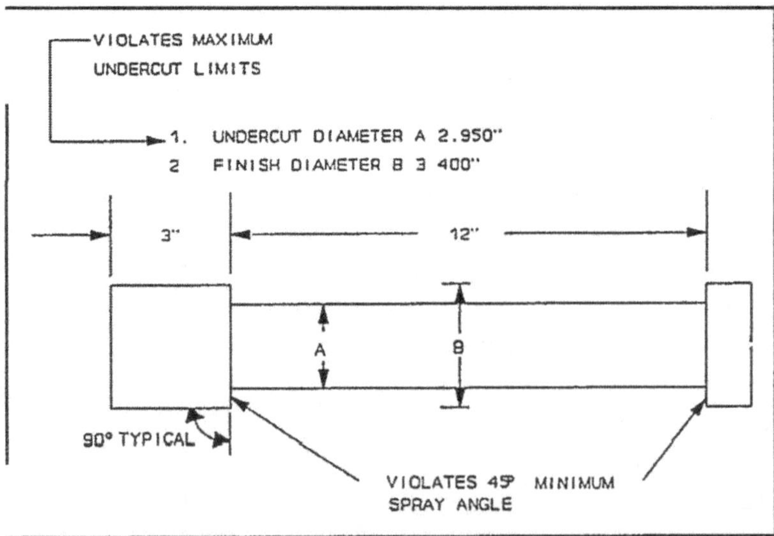

Figure 6.4 POORLY DESIGNED UNDERCUT

TABLE 6.1		
RECOMMENDED MAXIMUM DEPTHS OF UNDERCUTTING		
DIAMETER OF SHAFT	CLASS A SERVICE	CLASS B SERVICE
UNDER 1“	2.5% OF DIA.	4% OF DIA.
111-211	0.025”	0.040”
2" - 4"	0.035”	0.050”
4" - 6"	0.045”	0.060”
OVER 6“	0.055”	0.070”

CLASS A SERVICE For shafts or journals operating under a heavy bearing load or subject to severe service and high pressurre; applications where the maximum safety is required.

CLASS B SERVICE For normal duty lubricated bearing service; applications where a minimal safety factor is required.

TABLE 6.2		
NORMAL CONDITION COATING LIMITS		
SPRAY MATERIAL	PROCESS1/	MAXIMUM COATING THICKNESS IN NCHES)2/
CARBON STEEL -WIRE	AW	0.050
AUSTENITIC STAINLESS (WITH BOND COAT) -WIRE	FW	0.050
420 STAINLESS -POWDER -WIRE -WIRE	 PP AW FW	 0.075 0.050 0.050
NICKEL—ALUMINUM -POWDER	FP, PP	0.050
ALUMINUM-BRONZE -POWDER -WIRE	 FP, PP FW	 0.125 0.125
NICKEL-COPPER -POWDER -WIRE	 FP, PP AW	 0.040 0.040
COPPER-NICKEL -POWDER -WIRE	 FP, PP AW	 0.040 0.040
ALUMINUM-TITANIA -CERAMIC POWDER	FP, PP	0.015
BABBITT -POWDER -WIRE (WITH BOND COAT)	FP, AW	NO LIMIT

(From MIL-STD-1687)

1/ Spray processes are arc-wire (AW), flame-wire (FW), plasma-powder (PP), and flame powder (FP), applied as single or dual coating systems. Dual coating systems are identified by the bond coat material and finish coat material.

2/ The maximum coating thickness that can be successfully applied depends on the specific component dimensions, the specific chemistry of the spray material, and other factors.

UNDERCUT DESIGN FOR SPECIAL CONDITIONS

1. Occasionally, damage to a shaft may extend past the normal length of the undercut limits. If so, one or more of the following guidelines will apply:
   a. Special masking procedures may permit the undercut to extend to 1/8" from smfaces which must be protected from grit blast damage or 1/16" from end of shaft.
   b. If damage extends to the end of the shaft, and if no mechanical or metallurgical damage will be caused by welding, a weld bead may be deposited around the end of the repair area to provide the 1/16" minimum shoulder needed for thermal spraying.
   c. If valve stem backseat design permits minor reduction in its length during finish grinding, then the undercut may extend to the edge of the backseat (See Figure 6.5).

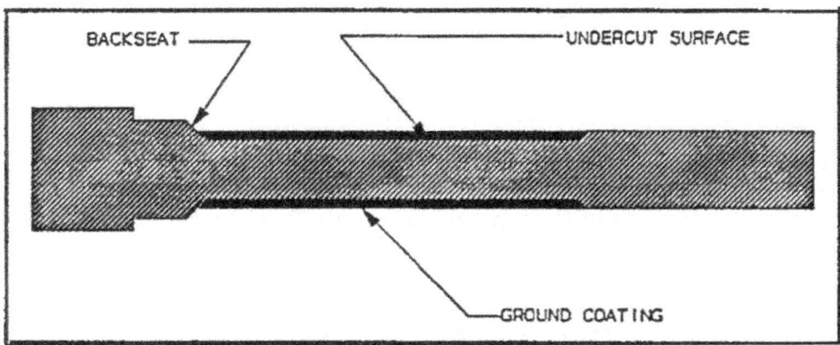

Figure 6.5 SPECIAL UNDERCUT DESIGN ON A VALVE STEM

   d. When weld buildup cannot be performed around the end of a shaft, a chamfer step may be cut at the end of the shaft (See Figure 6.6).

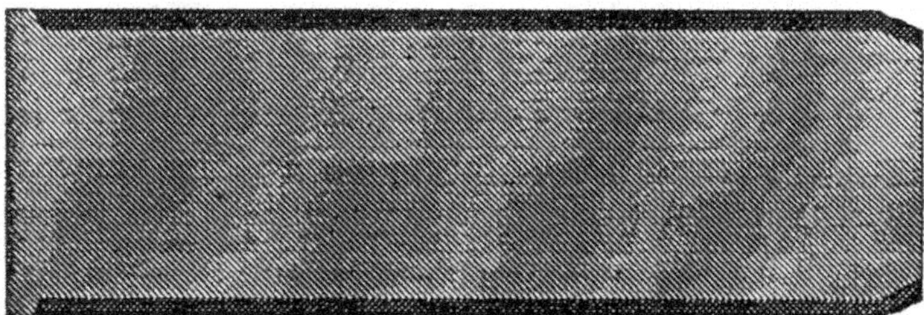

Figure ***6.6*** **UNDERCUT DESIGN WITH CHAMFER**

2. Depth of undercut limits for TABLE 6.1 may be exceeded only under the following conditions.
   a. Limits of TABLE 6.1 may only be exceeded if conditions of both b. and c. (below) are met.
   b. Undercut does not exceed either limits of thermal spray material manufacturers recommendations or limits of service/mock-up testing experience.
   c. Engineers determine that undercut does not affect service requirements, such as shaft strength due to the geometry of the component (See Figure 6.7).
   d. The limits of TABLE 6.2 tend to be conservative due to factors listed in Note 2 of that table. If surface damage is deeper than the limits of TABLE 6.2, consult thermal spray material manufacturers for guidance.

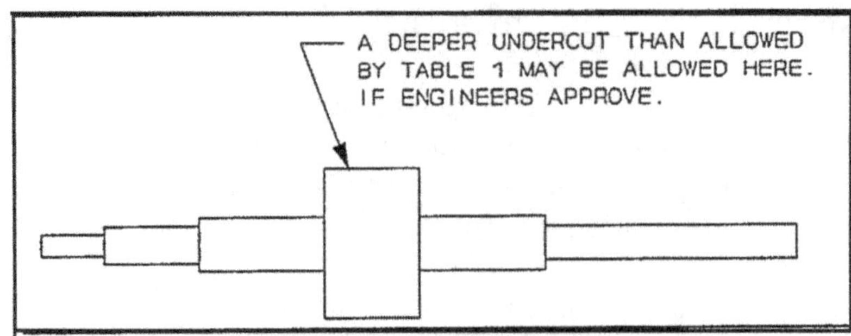

Figure 6.7 ABNORMAL UNDERCUT DESIGN

3. From time to time, shafts will be received in an undersize condition. These may sometimes be sprayed without undercut if remaining spray material after finish machining is at least as thick as the minimum requirements of the procedure and application (See normal guidelines for depth of undercut, TABLES 6.1 AND 6.2). If possible, a coating thickness of .010" per side should be applied. A minimum thermal sprayed coating of .005" per side is recommended for machinery components. If the coating is applied to thin, the thermal sprayed coating may overheat and cause spalling from the substrate.
4. A hole in the middle of a spray area may be treated in one of three ways..
   Interrupt the spray zone (See Figure 6.8).
   a Treat the hole in a manner similar to the spraying of keyways by using inserts.
   c. If the function of the hole is not affected by the spray process, the hole could be ignored.
5. Extremely heavy buildup (in excess of. 125") should not be attempted without years of experience. and practical experience on mock-ups and throw-away shafts.

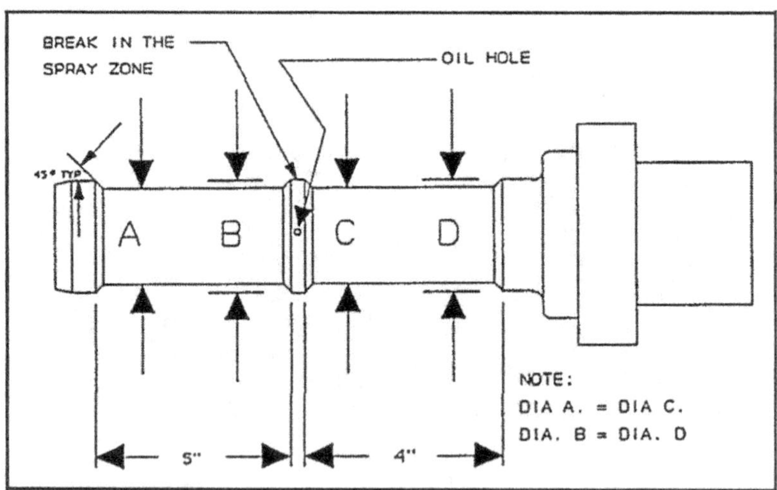

Figure 6.8 UNDERCUT DESIGN FOR OIL HOLE (Interrupted Spray Zone)

INTERNAL **DIAMETERS:**

Basically, the same recommendations apply (both to internal and external diameters), except that maximum thickness must be reduced for inside diameters due to the contraction of the sprayed material. This contraction process tends to shrink the coating away from the substrate, placing high tensile stresses on the bond interface (See Figure 6.9). For some internal diameters, special anchor-tooth blasting, thermal spray, and finishing equipment will be required. Many blasting, thermal spray, and machine equipment manufacturers market such equipment.

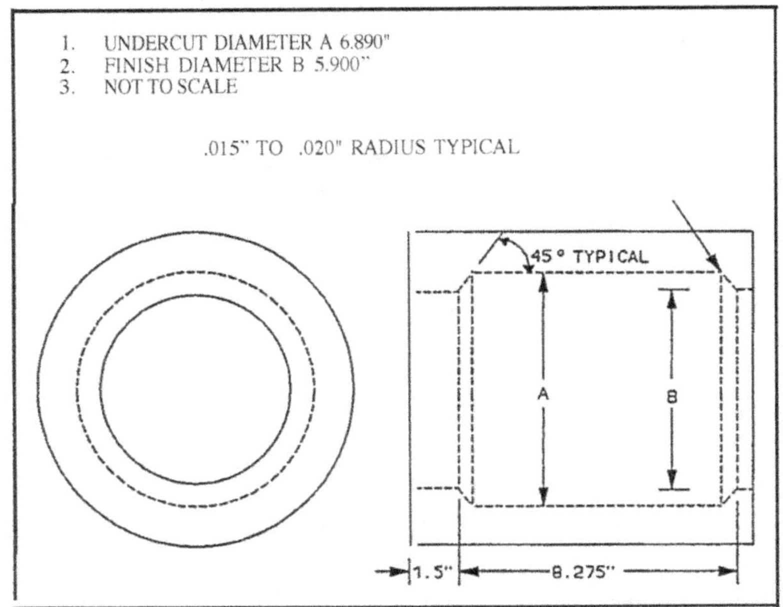

Figure 6.9 EXAMPLE OF AN INSIDE DIAMETER UNDERCUT DESIGN

MASKING

Critical machined areas which are not to be coated and would otherwise be damaged by grit blasting or thermal spraying must be masked. The thermal spray industry uses numerous masking methods and materials in the protection of these critical areas. Masking methods include covering, coating, and shadowing. Masking materials that may be used include tape, pipe, sheetmetal, masking compound, dummy keys, rubber, carbon, etc.

NOTE: Materials which cause corrosion or contamination of the thermal sprayed coating or substrate must not be used.

Almost without exception thermal spray masking must provide two types as well as two levels of protection. For most components, the masking within 4" to 6" of the area to be sprayed provides both blast and spray protection. This means that the masking must withstand the destructive forces of the blast stream as well as the direct heat of the spray stream. Masking which extends beyond this zone may receive some relatively minor heat stress, and it will receive some amount of impact from ricocheting blast media. When applying masking, operators must constantly consider the level, the type, and the direction of stress the masking material will receive. This consideration will help to determine the type and the amount of masking material applied. It will also help to determine the design of re-usable masking fixtures and materials.

TAPES

Several kinds of heat-resistant tape can be used as masking for grit blasting and can be left in place for thermal spraying. These tapes are easy to use and can offer good protection in most machinery component thermal spray applications. After grit blasting, the tape should be inspected to insure its condition is still adequate for use as thermal spray masking.

If non heat-resistant tape is used to mask for grit blasting, it must be removed and a heat resistant masking suitable for thermal spraying must be utilized. When applying the tape, it is advisable to use at least two layers, even if a shadow mask is used over it. If tape is the only masking used, three or more layers in the first four inches next to the spray zone may be necessary for adequate protection. After applying each layer, the air bubbles should be removed by firmly rubbing the tape. Inspection of the tape should take place from time to time during blasting and thermal spraying. Tape should be replaced if its ability to protect the critical areas of the machinery component is diminished. Large surfaces may be covered by sheets of light material (cloth, rubber, plastic, etc.) to save tape. The type of material depends on the proximity to spray and blast areas.

MASKING COMPOUNDS

Liquid masking compounds may be used to protect critical areas during the thermal spraying process. These compounds may not be reliable for masking during grit blasting. The compounds may be applied by brushing, spraying, or dipping. When applying the compound, caution should be used to insure it does not get on the area to be thermal sprayed. Many of these compounds are water soluble for easy removal after spraying.

SHADOW MASKING

Because it is usually desirable that the coating edge not end sharply, a technique called shadow masking is frequently used (See Figure (6.10). With this masking method, the area to be protected is partially shielded by raising the mask a little distance (usually, approximately 1/8") away from the component's surface. This causes some of the buildup to taper from the edge of the spray zone to a short distance under the shadow mask. Masking collars must be large enough so that any the anchor-tooth blasting grit can fall out before thermal spraying begins. Otherwise, during spraying grit might become wedged between the collar and the component, causing removal problems during the de-masking operation, and possibly even scarring the component's surface. Because of the nature of the blast and spray environments, shadow masks should be placed over a double layer of tape or masking compound. This helps to prevent blast damage and speeds cleanup and demasking after thermal spraying has been completed.

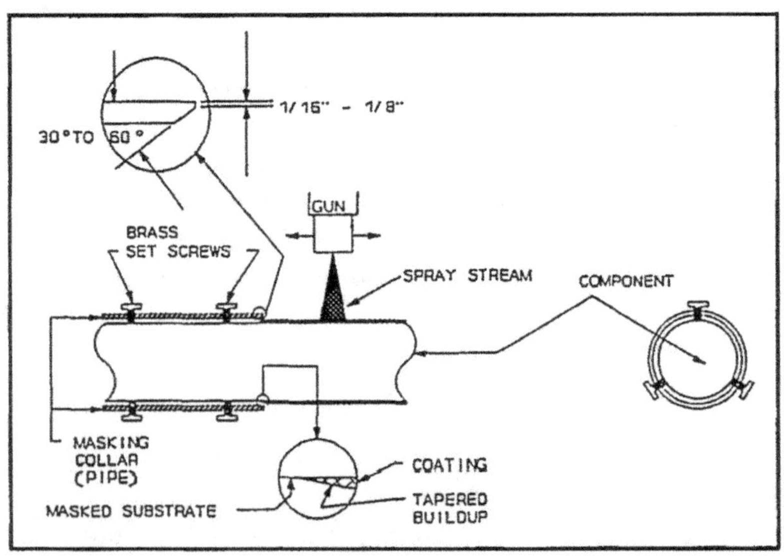

INSERTS

Some components include recessed areas, such as slots, holes, or keyways. These recessed areas cannot be effectively protected by tape or shadow masks. For these areas, inserts must be made (See Figure 6.11 for keyway insert design.). Materials for inserts vary according to the circumstances, and may include steel, aluminum, brass, copper, rubber, or carbon. Most metals are good for blasting protection and can be re-used many times for this purpose. These metal inserts are usually difficult to remove if left in for spraying. Rubber is only useful for blasting protection. Carbon can only be used during the spray operation. The best procedure for most applications is to make a carbon insert to snugly fit the area and then set it aside in a clean safe place. After this, make a metal insert and firmly seat it in recessed area of the component. After the anchor tooth blasting has been completed, remove the metal insert and replace it with the previously made carbon insert.

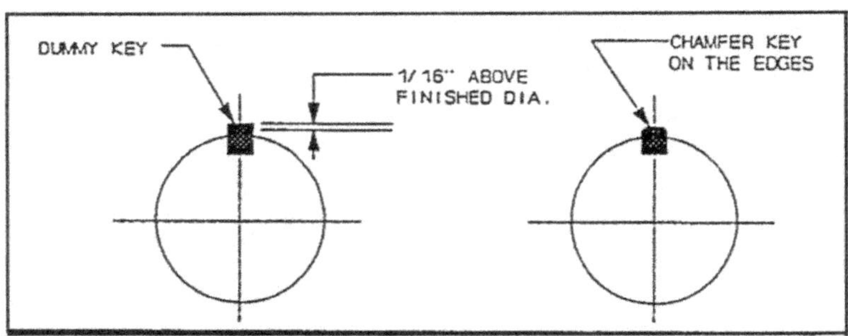

Figure 6.11 DUMMY KEY INSERT

Due to the widely varying nature of the recessed areas, specific instructions cannot be provided for all situations. However, the following guidelines, hints, and "Tricks-Of-The-Trade" will help operators to solve the unusual problems.

1. Inserts for spray protection should protrude no more than 1/16" above the finished surface of the component. This minimizes the shadowing effects of the insert on the coating in the adjacent area. Chamfering the upper edges of the insert will reduce the shadowing effect even more (See Figure 6.11).

NOTE Even when chamfered, the keyway mask partially shadows the edge of the keyway due to the rotation of the shaft.

2. Inserts for blast protection are restricted by two factors; they must survive the blast operation, and they must not prevent a good blast job in the adjacent area.
3. Carbon is usually by far the best material for spray protection because it is easily chipped out of the component after final machining without damage to the adjacent coating. However, it will not survive grit blasting. An insert of a different material must be made for that operation.
4. Inserts must fit snugly in order to prevent the blast or spray stream from blowing them out during initial passes.

NOTE: Inserts rarely fall out after the first pass. Blast or spray media tend to either seat them better or else knock them loose.

5. Steel keystock, if readily available, usually is best for blasting protection of slots and keyways. It generally fits snugly and requires minimal handwork.
6. For odd shaped or sized areas, soft metal is usually the best choice. Soft metal helps prevent scoring on the component. Aluminum is usually readily available, and is very easy to work with. It is easy to hammer, saw, or file. It survives blasting well enough to be re-used a number of times. Aluminum has a high enough melting point to withstand most spray temperatures if it must be left in place for spraying.
7. Machined grooves around the circumference may be masked by tying several wraps of heavy wire or by forming a rod into a temporary snap ring.
8. Snap ring grooves or slots might be masked by an old throw-away snap ring.
9. Material for masking fixtures can generally be acquired without purchasing new material by utilizing the following types of sources:
   a. Pipe, plate, or sheet metal from scrap bins or excess material racks.
   b. Odd shaped pieces from scrap bins.
   c. Throw-away parts removed during disassembly of the component to be thermal sprayed.
   d. Carbon rod readily available in most welding shops.

STAGING

To minimize the time between the surface preparation and the thermal spraying of the component, the thermal spray equipment, measuring equipment (micrometer, contact pyrometer, etc.), and the turning fixture should be set up before surface preparation is started. All tools that will touch the area to be sprayed after final surface preparation must be cleaned of all oil and grease. At this time the thermal spray equipment should also be set up as described in Section 7 of this manual. If the component is to be preheated, the thermal spiny equipment and turning fixture are set up to the same parameters that will be used to apply the thermal sprayed bond coat. The details of the staging operation are discussed in Section 8 of this manual in the portion prior to "Preheating".

ANCHOR TOOTH BLASTING

Final surface preparation for thermal spraying machinery components is the most important step of the entire coating procedure. Surface roughness of the area to be thermal sprayed has a direct influence on how well the coating adheres to the substrate material.

After masking is completed (just before anchor tooth blasting begins) the spray area must be solvent cleaned one final time. This step insures removal of any contaminants (including finger prints) which may have inadvertently been introduced during the masking process. Except for unusual conditions, this final cleaning is best done using a clean disposable towel which has been permeated with clean solvent. After all contaminants have been removed, the substrate should be protected from becoming re-contaminated. Depending on the circumstances, this protection can range from careful handling to wrapping with an inner layer of clean dry paper or clean cloth or an outer layer of rubber or plastic.

Anchor-tooth blasting with aluminum oxide grit (16-30 mesh) is the most commonly used method to complete the final surface preparation of machinery components to be thermal sprayed. This surface preparation method creates an anchor tooth pattern for the coating to adhere to (See Figure 6.12). To help insure the use of clean and correctly sized aluminum oxide, many facilities do not recycle grit in this operation. Instead, they move the expended grit to a different blast unit to be recycled for non-anchor tooth operations, such as corrosion removal or cleaning/descaling purposes. Thermal spray facilities typically use a dedicated blasting cabinet to complete anchor tooth blasting. A separate blasting cabinet should be used for cleaning operations. This helps to prevent oil contamination of surfaces which have already been cleaned and are receiving final anchor tooth blasting.

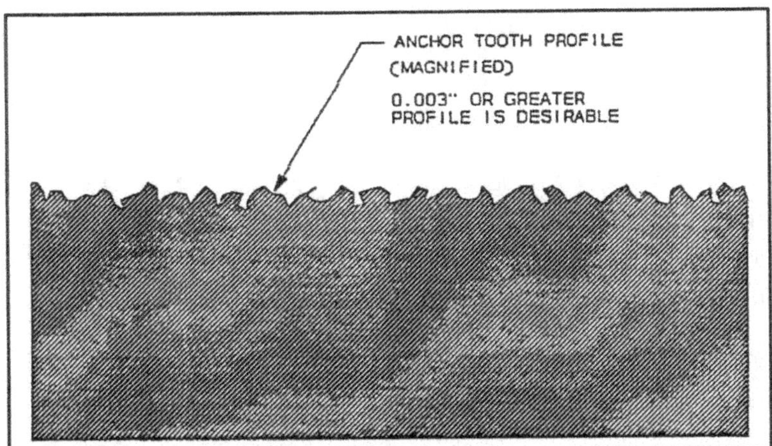

Figure 6.12 ANCHOR TOOTH PROFILE

The following is a list of blasting related equipment recommended for performing all blasting operations associated with the thermal spray operation.

1. Compressed air is limited to a maximum of 5 milligrams of condensed hydrocarbons per cubic meter and a dew point of + 14°F or lower at standard temperature and pressure (68"F, 14.5 lb/in^{2}absolute) prior to the final faltering and moisture separation unit. For abrasive blasting, minimum air pressure at the blast generator of 50 lb/in^{2}and 75 lb/in^2 is required for pressure type and suction type blasting units, respectively.
2. Air pressure regulators with pressure gauges.
3. Steel blasting table and cabinet.
4. Oil and moisture separator.
5. Blast nozzle equipped with a dead-man switch.
6. Aluminum oxide grit.
7. Profile tape.
8. Dial micrometer.

The Military Standard, "Thermal Spray Processes for Naval Ship Machinery Applications" (MIL-STD-1687), requires a 2 to 4 roils anchor-tooth surface profile for final surface preparation. This Standard also states that when distortion may be encountered due to part configuration, the anchor-tooth pattern may be reduced to a 1 mil profile (minimum) provided that a procedure qualification with the 1 mil profile meets all other MIL-STD-1687 requirements. Lower blasting pressure or greater nozzle-to-substrate distance can also be an effective method for reducing or eliminating distortion.

A good method for measuring the anchor tooth pattern of a component is with the use of profile tape and dial micrometer.

Visual inspection of the anchor-tooth profile shall assure that surface appearance is uniform. Steel substrates shall have a white metal blast finish, defined as a gray-white, uniform metallic color. Oil, grease, scale, paint, or any contaminant shall not be present on the substrate to be thermal sprayed after final surface preparation. After the final surface preparation, the area to be sprayed must not come in contact with any item which would contaminate the surface. Areas to be thermal sprayed should be handled with clean material. Even the oil from fingerprints may cause a coating failure. The slightest presence of oil, oxidation or other foreign material on the surface to be sprayed may result in separation of the thermal spray coating. The blasted surface is rejectable if the white metal blast condition is lost.

CONCLUSION

Producing a quality thermal spray coating requires a high degree of control over all stages of operations in this Section. The steps listed in this Section are critical to the success in creating a high quality thermal spray coating. It is essential during these steps for thermal spray personnel to understand the criticality of the cleanliness requirements.

8 MORE ON NAVAL APPLICATIONS

THERMAL SPRAY APPLICATIONS

INTRODUCTION

Thermal spraying is typically performed to restore the dimensions of a machinery component and to a lesser degree to improve surface performance. Some reasons for dimensional restoration include wear, erosion, corrosion or mismachining. Most of the machinery components repaired by this process are shafts. Some ship building and ship repair facilities have used the thermal spray process to repair inside bores. Other thermal spray facilities have repaired mating surfaces of pump casings and babbitt bearings.

HISTORY

Before the Naval Sea Systems Command would permit the use of the thermal spray process on shipboard components, testing was completed to determine the reliability and effectiveness of the process as a repair method for critical machinery components. The Naval Sea System Command; the Naval Surface Warfare Center Annapolis, Maryland (formerly David Taylor Naval Ship Research and Development Center); and Puget Sound Naval Shipyard developed a program to assess the performance of new thermal spray coatings and techniques and to review past industrial applications. To determine the reliability of ships machinery thermal spray applications, the following four-pronged approach was developed:

1. Compile and evaluate case histories of thermal sprayed components.
2. Perform thermal spray coating laboratory and procedure qualification testing.
3. Perform land based service testing of thermal sprayed components.
4. Perform fatigue, wear, corrosion, and shock testing of thermal spray machinery components.

CURRENT APPLICATIONS

Because the above. four-pronged program was highly successful, the Naval Sea System Command approved thermal spray applications on many critical machinery components. The Naval Sea System Command's Military Standard 1687 covers the thermal spray processes for repair and overhaul of naval ship machinery and ordnance components. Permitted applications on surface ships presently include main feed pumps, forced draft blowers, and other equipment of equal or lesser criticality. One good example of a Naval Sea Systems Command approved application is the thermal spraying of 410 stainless steel steam valve stems. Steam valve stems have been approved for thermal spraying, because of the cost savings obtained (approximately $1,000.00 per stem) by repairing the stems instead of purchasing or manufacturing new ones. The coating system used to repair these components is a plasma sprayed bond coating of Eutectic 21031 (Nickel, Chromium, Molybdenum & Aluminum) and a plasma sprayed ceramic final coat of Metco 143 (Zirconia, Titania & Yttria). Frequently these thermal sprayed coatings perform better than new components. Thermal sprayed ceramic stems have superior stagnant fresh water and sea-water corrosion resistance.

Other approved repairs on machinery components include:

1. Repair of static fit areas to restore orginal dimensions, finish, and alignment.
2. Repair of seal (including packing) areas to restore original dimensions and finish.
3. Repair of fit areas on shafts to restore original dimensions and finish (except for motor generator sets).
4. Buildup on pump shaft sleeves and wear rings to restore original dimensions.
5. Repair of bearing shaft journals to restore original dimensions.
6. Babbitt bearings for auxiliary equipment.
7. Turbine pump casing flanges to restore original dimensions.
8. Ceramic for corrosion/erosion protection.
9. Exposed areas on shafts to restore surface finish requirements.
10. Ceramic to overcome galling tendencies of current designs.

On United States Navy Ships, applications not specified in Military Standard 1687 require Naval Sea Systems Command approval for each item to be sprayed. That approval should be based on the results of laboratory or service tests to determine the adequacy of using the thermal spray process in the proposed application.

TABLE 3.1 has been created to list some of the types of components which have been repaired by using the therrmal spray process. This list of components was obtained from the data base that is kept by the Naval Surface Warfare Center in Annapolis, Maryland. All of the machinery components listed are known successful thermal spray applications performed by Puget Sound Naval Shipyard, Naval Intermediate Maintenance Activities (IMAs), the Naval Shore Repair Facility (SRI) Guam, and private ship repair facilities.

CONCLUSION

In addition to the various types of components currently in service, the potential exists for many new thermal spray applications on military and commercial vessels. Manufacturers produce a wide variety of thermal spray materials to meet the needs of many applications. TABLE 3.2 list a broad range of current and potential applications for which coating materials exist. It is important to note that many of the applications suggested by this table go far beyond the current limits of MIL-STD-1687 approval.

TABLE 3.1

CURRENT APPLICATIONS

COMPONENT	SHIPS SYSTEM	AREA REPAIRED	**BASE MATERIAL**	ENVIRONMENT	PROCESS *	THERMAL SPRAY MATERI
FORCE DRAFT BLOWER SHAFT	FORCE DRAFT BLOWER	LABYRINTH SEAL	CARBON STEEL	STEAM	FINAL PP	EUTECTIC 21031
PUMP SHAFT	CONTAMINATED HOLDING TANK	PACKING	NiCu	SALT WATER	BOND PP	DRESSER PP-25
					FINAL PP	METCO 130
WORM SHAFT	LUBE OIL	SEAL	CARBON STEEL	FUEL OIL	FINAL PP	METCO 444
ROTOR SHAFT	PUMP MOTOR	JOURNAL	NiCu	FRESH WATER	FIINAL PP	DRESSER P-25
VALVE STEM	MAIN STOP VALVE	PACKING	410 STAINLESS STEEL	STEAM	BOND PP	EUTECTIC 21031
					FINAL PP	METCO 130
SHAFT SLEEVE	CIRCULATING PUMP	PACKING	K-MONEL	SALT WATER	BOND PP	DRESSER PP-25
					FINAL PP	METCO 130
VALVE STEM	MS-9 VALVE	PACKING	410 STAINLESS	STEAM	BOND PP	EUTECTIC 21031
					FINAL PP	METCO 143
PUMP SHAFT	SALT WATER COOLING	FIT	NiCu	SALT WATER	FINAL PP	DRESSER PP-25
VALVE FLANGE	RELIEF VALVE	FLANGE	BRONZE	SALT WATER	FINAL FW	METCO SPRAYBRONZE AA
PUMP SHAFT	DISTILLATION	JOURNAL	NiCu	FRESH WATER	FINAL FP	DRESSER PP-25

* **AW = ARC WIRE PROCESS**
FW = FLAME WIRE PROCESS
PP = PLASMA POWDER PROCESS
FP = FLAME POWDER PROCESS

TABLE 3.1

CURRENT APPLICATIONS

COMPONENT	SHIPS SYSTEM	AREA REPAIRED	BASE MATERIAL	ENVIRONMENT	PROCESS	THERMAL SPRAY MATERIAL
END BELL	PUMP MOTOR	EDGE	CAST IRON	SALT WATER	FINAL PP	EUTECTIC 21031
MOTOR SHAFT	AC MOTOR	FIT	CARBON STEEL	AIR	FINAL PP	METCO 444
PUMP SHAFT	LUBE OIL	SEAL	CARBON STEEL	LUBE OIL	FINAL PP	METCO 443
SHAFT	MAIN FEED BOOSTER	SEAL	CARBON STEEL	SALT WATER	BOND PP	EUTECTIC 21031
					FINAL PP	METCO 130
STRAINER PLUG	SALT WATER CIRCULATION	OUTSIDE DIAMETER	BRONZE	SALT WATER	FINAL AW	METCO SPRAYBRONZE AA
STRAINER PLUG	SALT WATER CIRCULATION	OUTSIDE DIAMETER	BRONZE	SALT WATER	FINAL FW	METCO SPRAYBRONZE AA
STRAINER PLUG	LUBE OIL	OUTSIDE DIAMETER	BRONZE	LUBE OIL	FINAL FW	METCO SPRAYBRONZE AA
TURBINE ROTOR SHAFT	AUXILIARY STEAM	PACKING	CARBON STEEL	STEAM	BOND PP	METCO 443
					FINAL PP	METCO 130
TURBINE ROTOR SHAFT	AUXILIARY STEAM	BEARING JOURNAL	CARBON STEEL	OIL	BOND PP	METCO 443
					FINAL PP	METCO 130

* AW = ARC WIRE PROCESS
FW = FLAME WIRE PROCESS
PP = PLASMA POWDER PROCESS
FP = FLAME POWDER PROCESS

TABLE 3.1

CURRENT APPLICATIONS

COMPONENT	SHIPS SYSTEM	AREA REPAIRED	BASE MATERIAL	ENVIRONMENT	* PROCESS	THERMAL SPRAY MATERIAL
END BELL	VENTILATION SYSTEM	BEARING JOURNAL	CARBON STEEL	AIR	FINAL FP	METCO 444
PLUG	DUPLEX STRAINER	OUTSIDE DIAMETER	BRASS	SALT WATER	BOND FW	METCO 405
					FINAL FW	METCO SPRAYBRONZE AA
STUB SHAFT	MAIN PROPULSION	SEAL	BRONZE	SALT WATER	BOND PP	METCO 445
					FINAL PP	METCO 130
DONKEY BOILER	RECIRCULATING PUMP	PACKING	CARBON STEEL	CONDENSATE STEAM	FINAL PP	EUTECTIC 21031
VALVE STEM	MAIN FEED PUMP GOVERNOR	PACKING	STAINLESS STEEL	STEAM	BOND PP	EUTECTIC 21031
					FINAL PP	METCO 143
ROTOR SHAFT	CONTAMINATED HOLDING TANK	IMPELLER FIT	NiCu	SEWAGE/ WASTE	FINAL PP	DRESSER PP-25
PUMP PLUNGER	BRINE PUMP	PACKING	STAINLESS STEEL	BRINE	BOND PP	METCO 444
					FINAL PP	METCO 136F
BEARING HOUSING	BALLAST PUMP	INSIDE DIAMETER	BRONZE	GREASE	FINAL PP	METCO 445

* AW = ARC WIRE PROCESS
FW = FLAME WIRE PROCESS
PP = PLASMA POWDER PROCESS
FP = FLAME POWDER PROCESS

TABLE 3.1

CURRENT APPLICATIONS

COMPONENT	SHIPS SYSTEM	AREA REPAIRED	BASE MATERIAL	ENVIRONMENT	PROCESS *	THERMAL SPRAY MATERIAL
BABBITT BEARING SHELL	LUBE OIL PUMP	OUTSIDE DIAMETER	CARBON STEEL	AIR	FINAL PP	METCO 447
EXPANSION JOINT	CATAPULT STEAM	PACKING	CuNi	STEAM	BOND PP	EUTECTIC 21021
					FINAL PP	METCO 130
BABBITT BEARING	TURBINE GENERATOR	INSIDE DIAMETER	BABBITT	OIL	FINAL AW	BABBITT
PLUNGER	AIRCRAFT ELEVATOR	OUTSIDE DIAMETER	CARBON STEEL	GREASE	FINAL AW	METCO METCOLOY #2
FLOW SERVICE PUMP	FUEL OIL	PACKING	CARBON STEEL	FUEL OIL	BOND PP	METCO 447
					FINAL PP	METCO 130
POTABLE WATER MOTOR SHAFT	POTABLE WATER	JOURNAL	NiCu	POTABLE WATER	FINAL PP	DRESSER PP-25
ROTOR SHAFT	WATER PUMP	PACKING	CARBON STEEL	FRESH WATER	BOND PP	METCO 447
					FINAL PP	METCO 130
MOTOR SHAFT	MOGAS PUMP	FIT	CARBON STEEL	AIR	FINAL PP	METCO 444
ROTOR SHAFT	SALT WATER COOLING PUMP	PACKING	NiCu	SALT WATER	BOND PP	DRESSER PP-25
					FINAL PP	METCO 130

* AW = ARC WIRE PROCESS
FW = FLAME WIRE PROCESS
PP = PLASMA POWDER PROCESS
FP = FLAME POWDER PROCESS

TABLE 3.1

CURRENT APPLICATIONS

COMPONENT	SHIPS SYSTEM	AREA REPAIRED	BASE MATERIAL	ENVIRONMENT	* PROCESS	THERMAL SPRAY MATERIAL
ROTOR SHAFT	VENTILATION	FIT	CARBON STEEL	AIR	FINAL PP	METCO 447
PUMP SHAFT	SSTG PUMP	FIT	CARBON STEEL	AIR	FINAL PP	METCO 444
HD-977 SP PUMP SHAFT	SP-5-49 COOLING	SEAL	NiCu	SALT WATER	BOND PP	DRESSER PP-25
					FINAL PP	METCO 130
PUMP SHAFT	ELECTRONIC COOLING	SEAL	NiCu	FRESH WATER	BOND PP	DRESSER PP-25
					FINAL PP	METCO 130
PUMP CASING	FIRE PUMP	INSIDE DIAMETER (BEARING JOURNAL FIT)	CARBON STEEL	OIL	FINAL AW	TAFA 75B
PUMP SHAFT	CHILL WATER PUMP	FIT	NiCu	AIR	FINAL AW	EN-60
THRUST ROLLER PLUNGER	AIRCRAFT ELEVATOR		CARBON STEEL	GREASE	FINAL AW	METCO METCOLOY #2
PLUNGER	AIRCRAFT ELEVATOR	SEAL	CARBON STEEL	GREASE	BOND PP	METCO 444
					FINAL PP	METCO 130

* AW = ARC WIRE PROCESS
FW = FLAME WIRE PROCESS
PP = PLASMA POWDER PROCESS
FP = FLAME POWDER PROCESS

TABLE 3.1

CURRENT APPLICATIONS

COMPONENT	SHIPS SYSTEM	AREA REPAIRED	BASE MATERIAL	ENVIRONMENT	* PROCESS	THERMAL SPRAY MATERIAL
PISTON ROD	AIRCRAFT ELEVATOR	SEAL	CARBON STEEL	AIR	BOND PP	METCO 447
					FINAL PP	METCO 130
STEAM VALVE STEM	CATAPULT	PACKING	410 STAINLESS STEEL	STEAM	BOND PP	METCO 444
					FINAL PP	METCO 143
IMPELLER	CATAPULT	FIT	GUN METAL BRONZE		FINAL AW	AMPCO 10
PUMP SHAFT	EMERGENCY DIESEL	PACKING	STAINLESS STEEL		BOND PP	METCO 447
					FINAL PP	METCO 130
ROTOR SHAFT	DEFUELING PUMP	FIT	NiCu	OIL	FINAL AW	EN-60
RETRACT SHEAVE SHAFT	ARRESTING GEAR	FIT	CARBON STEEL	STEAM	FINAL PP	METCO 447

AW = ARC WIRE PROCESS
FW = FLAME WIRE PROCESS
PP = PLASMA POWDER PROCESS
FP = FLAME POWDER PROCESS

TABLE 3.2	
POTENTIAL THERMAL SPRAY APPLICATIONS	
GENERAL APPLICATIONS	SPECIFIC CONDITIONS
WEAR RESISTANCE	SOFT BEARINGS HARD BEARINGS ABRASIVE RESISTANT FIBER/THREAD CAVITATION PARTICLE EROSION
HEAT & OXIDATION RESISTANCE	OXIDIZING ATMOSPHERES CORROSIVE GASES THERMAL BARRIERS MOLTEN METAL HANDLING
CORROSION RESISTANCE	INDUSTRIAL & SALT ATMOSPHERE FRESH WATER SALT WATER CHEMICAL & FOOD PROCESSING
ELECTRICAL	CONDUCTIVE RESISTORS INSULATORS RADIO FREQUENCY SHIELDING
RESTORATION OF DIMENSIONS	MACHINABLE MATERIALS GRINDABLE MATERIALS VARIOUS ALLOYS
MACHINE ELEMENT CLEARANCE CONTROL	ABRADABLE COATINGS ABRASIVE COATINGS
CHEMICAL CORROSION RESISTANCE	METAL COATINGS CERAMIC COATINGS CERMETS PLASTICS

The thermal spray materials tabulated in this manual have been so listed because of the experience Naval facilities have had using these manufacturers in the past. It is to be noted that other companies also produce excellent quality and equivalent thermal spray materials.

GLOSSARY

ABRASIVE Material used for cleaning and/or for final surface preparation for applying a thermal sprayed coating. For the final surface preparation of machinery components, aluminum oxide is normally used.

ACETYLENE A fuel gas that can be used in the flame wire and flame powder thermal spray processes.

ADHESION A binding force that holds together molecules of substances whose surfaces are in contact or near proximity.

ALUMINA: A chemical compound (aluminum oxide); a ceramic used in powder or rod form in thermal spraying operations. May also be a blasting medium.

ANCHOR-TOOTH PROFILE The condition of the substrate after abrasive blasting for surface roughening.

ARC WIRE PROCESS: A thermal spray process where a heating zone is created by an arc between two continuously fed metallic wires, one positively charged and one negatively charged. The heating zone melts the wire material and compressed air or an inert gas is used to propelled the molten material to the substrate.

ARC: A luminous discharge of electrical current crossing the gap between two electrodes.

ARGON: An inert gas that can be used in plasma powder thermal spray process.

BASE MATERIAL The material of the component being thermal sprayed.

BLASTING: A method of cleaning or surface roughening by a forcibly projected stream of sharp angular abrasive.

BOND COAT A preliminary (or prime coat) of material that improves adherence of the subsequent spray deposit.

BOND STRENGTH: The force required to pull a coating free of a substrate, usually expressed in psi (kPa).

CARRIER GAS: The gas used to carry powdered material from the powder feeder or hopper to the thermal spray gun.

CERAMICS: Various hard, brittle, heat and corrosion-resistance materials normally used as the final coat with the thermal spray process.

CERMET: A physical mixture of ceramics and metals; examples are alumina plus nickel and zirconia plus nickel.

COATING SYSTEM: One or more thermal spray coatings that are qualified for use singly or in combination. An example of a one coating system is a stainless steel coating applied for restoration of dimensions. An example of a two coating system is a nickel base bond coat and ceramic final coat.

COATING STRESS: The stresses in a coating resulting from rapid cooling of molten material or semimolten particles as they impact the substrate. Coating stresses area combination of body and textural stresses.

COATING: (1) The act of building a deposit on a substrate, (2) the spray deposit.

COEFFICIENT OF THERMAL EXPANSION: The ratio of the change in length per degree rise in temperature to the length at a standard temperature such as 68°F (20°C).

COMPOSITE: Each grain contains all chemical elements present in the powder, but they are not alloyed.

DEFECT: A discontinuity or group of discontinuities that by nature or accumulated effect (for example, total crack length) render a part or product unable to meet minimum applicable acceptance standards or specification. This term designates rejectability.

FINAL COAT: The last coating to be applied during the thermal spraying sequence.

FIXTURING: Mechanical positioning to support the machinery component and thermal spray gun.

FLAME WIRE PROCESS: A thermal spray process that uses an oxygen fuel flame to create a heating zone. A wire is then continuously fed into the heating zone where is melted and atomized and then propelled to the substrate by the force of the burning gasses and compressed air.

FLAME POWDER PROCESS: A thermal spray process that uses an oxygen fuel flame to create a heating zone. Powder is fed into the heating zone and melted. The molten material is then propelled to the substrate by the force of the burning gasses and compressed air.

FRETTING: Surface damage resulting from relative motion between surfaces in contact under pressure.

GRIT BLASTING: See Blasting.

HELIUM: An inert gas that can be used in the plasma powder thermal spray process.

HIGH VELOCITY OXYGEN FUEL PROCESS: A thermal spray process where oxygen and fuel are mixed at high pressures. This mixture is ejected from a nozzle and ignited.
Into the high velocity flame powder is injected. The powder is propelled to the substrate by the force of the burning gases. When the thermal spray material collide with the substrate it plasticizes and forms a coating.

HYDROGEN: A fuel gas that can be used with plasma powder and high velocity oxygen fuel thermal spray processes.

INERT GAS: A gas which does not normally combine chemically with the substrate or the deposit.

INTERFACE The contact surface between the spray deposit and the substrate and between the bond coat and final coat.

INTERPASS TEMPERATURE In multiple pass thermal spraying, the temperature (minimum or maximum as specified) of the deposited thermal spray coating before a subsequent pass is started.

MASKING: The method of protecting the areas adjacent to the area to be abrasive blasted and thermal sprayed.

MECHANICAL BOND: The principal bond that holds thermal spray materials together and is formed between base metals and filler metals in all processes.

METALLIZING: See preferred term "thermal spraying".

MICROMETER Any device for measuring minute distances, especially one based on the rotation of a freely threaded screw.

MUST The word "must" is to be understood as mandatory for the purpose of avoiding coating failures. *

NITROGEN: An almost inert gas that can be used with the plasma powder process.

NONTRANSFERRED ARC: An arc established between the electrode and the constricting nozzle. The workpiece is not in the electrical circuit.

NOZZLE (1) A device which directs shielding media, (2) a device that atomizes air in an arc wire spray gun, (3) the anode in a plasma gun, (4) the gas burning jet in a powder, rod, or wire flame spray gun.

OXIDES: A binary compound of an element or radical with oxygen.

OXYGEN A gas that is used with the flame wire, flame powder, and high velocity oxygen fuel thermal spray processes.

PARAMETER A measurable factor relating to several variables; loosely used to mean spraying variable, spraying condition, spraying distance, angle gas pressure, gas flow, etc.

PASS: A single progression of the thermal spray device across the surface of the substrate.

PLASMA POWDER PROCESS: A thermal spray process where a heating zone is created by an arc between a tungsten electrode and a nozzle. A gas or gas mixture is passed through this arc. This gas or gas mixture then exits the nozzle, forcing the flame outside the gun. A powder is fed into the flame where it is melted and propelled to the substrate by the force of the gas or gas mixture.

POROSITY: Cavity type discontinuities within a thermal sprayed coating.

PREHEAT: Heat applied to the substrate prior to thermal spraying. This may be performed to remove moisture and/or to reduce residual stresses.

PRIMARY GAS: The major constituent of the arc gas fed to the gun to produce the plasma; usually argon or nitrogen.

PROCEDURE The detailed elements of a process or method used to produce a specific result.

PYROMETER. An electrical thermometer for measuring high temperatures.

QUALITY ASSURANCE (QA): The activity of providing to all concerned the evidence needed to establish confidence that the quality function is being performed adequately (from the Juran Quality Control Handbook, 3rd edition).

SEALER: Material applied to thermal sprayed coatings to close pores and facilitate finishing.

SECONDARY GAS: The minor or second constituent of the arc gas fed to the gun to produce plasma.

SHADOW MASKING: A protective device that partially shields from a small distance above the work, thus permitting some overspray to produce a feathering at the coating edge.

SHOULD: The word "should" is to be understood as advisory for the purposes of improving coating quality and/or increasing the efficiency of the operation.

SHOULDER Area of component at the end of the undercut.

SOLVENT CLEANING: A method used to remove contaminants from the machinery component that is to be thermal sprayed.

SPALLING: The flaking or separation of a thermally sprayed coating.

SPLAT: A single thin flattened sprayed particle,

SPRAY ANGLE The angle of particle impingement, measured from the surface of the substrate to the axis of the spraying nozzle.

SPRAY DISTANCE The distance maintained between the thermal spraying gun nozzle tip and the surface of the component during spraying.

SPRAY RATE The rate at which surfacing material passes through the thermal spray gun.

SUBSTRATE Any material to which a thermally sprayed coating is applied.

SURFACE PREPARATION The operation necessary to produce a desired or specified surface condition.

SURFACE FEET PER MINUTE: The circumferential velocity of the substrate.

TENSILE BOND STRENGTH: See preferred term, "bond strength. ”

THERMAL SPRAYING: A group of processes where freely divided metallic or non-metallic materials are deposited in a molten or semi-molten condition to form a coating.

THERMAL SPRAY GUN A device for heating, feeding and directing the flow of a surfacing material.

TRAVERSE SPEED: The linear velocity at which the thermal spray gun traverses across the substrate during the spraying operation.

TSJCR Thermal Spray Job Control Record.

TURNING FIXTURE Mechanized equipment (i.e. lathe and traversing gun mount) that keeps the gun to-substrate distance and spray angle freed during spraying and the relative movement between the gun and substrate accurately controlled.

UNDERCUTTING: A step in the sequence of surface preparation involving the removal of substrate material.

WORKPIECE The object or surface to be coated. See preferred term substrate.

* Although guidelines using the word "must" can generally be shown to have exceptions, the experience of the authors has been that any exceptions are so minor in nature or frequency that they should be disregarded. (See definition of "should").

* Definitions of Thermal Spray terms can be found in the American Welding Society’s Thermal Spraying Manual “Practice, Theory, and Application”.

SECTION II

Engineering and Design
THERMAL SPRAYING: New Construction and Maintenance

Revised by Wexford Press
from
EM 1110-2-3401
U.S. Army Corps of Engineers

Engineering and Design
THERMAL SPRAYING: NEW CONSTRUCTION AND MAINTENANCE

Table of Contents

Appendix A
References

Appendix B
Glossary

Appendix C
Summary Description of CEGS-09971, "Metallizing: Hydraulic Structures"

Appendix D
USACE Field Experience and Lessons Learned

Subject Index

Chapter 1
Introduction

1-1. Purpose

This manual provides guidance on thermal spray coating systems to engineering, operations, maintenance, and construction personnel and other individuals responsible for the protection of U.S. Army Corps of Engineers (USACE) civil works structures. It gives broad-base instructions on corrosion protection using thermal sprayed coatings and state-of-the-art procedures that can be employed on Corps projects, which can aid in attaining better and, from a long-range viewpoint, more economical thermal spray jobs.

1-2. Applicability

This Engineer Manual (EM) applies to all USACE Commands having responsibilities for the design and construction of civil works projects.

1-3. References

Required and related references are listed in Appendix A.

1-4. Distribution Statement

Approved for public release; distribution is unlimited.

1-5. Abbreviations and Acronyms

A glossary containing abbreviations and acronyms used herein is included as Appendix B.

1-6. Neutral Language Use and Terms

a. Throughout this manual neutral language is used to include both the masculine and feminine genders: any exceptions to this statement will be noted.

b. The terms "metallize," "metallizing," and "thermal spraying" are used broadly herein to indicate all types of sprayed metal protective coatings applied by any of several processes. The term "thermal spraying" is more inclusive and is descriptive of other non-metallic as well as metallic coating materials.

1-7. Scope

a. This manual presents an overview of thermal spray technology and discusses thermal spray coating materials, processes, specifications, selection, surface preparation, application, inspection and testing, sealing, maintenance, and safety.

b. Certain types of thermal spray coatings and processes are discussed even though they do not appear in Guide Specification CEGS-09971. Some of these thermal spray coatings and processes are presented for general information; others are discussed because they may be considered for use in special situations.

c. Paint materials are discussed in this manual to the extent that this information is useful in understanding the applicability of thermal spray coatings. The inclusion of information on paint coatings illustrates that the subject of corrosion protection using coatings cannot be approached in an exclusive manner.

d. This manual includes a complete topic index to facilitate locating specific information. Appendix A will help locate specific references mentioned. Appendix B lists abbreviations and acronyms used. Appendix C contains important requirements needed for specifying a complete metallizing specification for a civil works project. Appendix D is provided to illustrate recent USACE experience and lessons learned with thermal spray coatings at civil works facilities.

Chapter 2
Thermal Spray Fundamentals

2-1. Introduction

This chapter introduces the engineer to the fundamental principles of thermal spray, coating types and characteristics, thermal spraying processes, and thermal spray uses.

2-2. General Description of Thermal Spraying

Thermal spraying is a group of processes wherein a feedstock material is heated and propelled as individual particles or droplets onto a surface. The thermal spray gun generates the necessary heat by using combustible gases or an electric arc. As the materials are heated, they are changed to a plastic or molten state and are confined and accelerated by a compressed gas stream to the substrate. The particles strike the substrate, flatten, and form thin platelets (splats) that conform and adhere to the irregularities of the prepared substrate and to each other. As the sprayed particles impinge upon the surface, they cool and build up, splat by splat, into a laminar structure forming the thermal spray coating. Figure 2-1 illustrates a typical coating cross section of the laminar structure of oxides and inclusions. The coating that is formed is not homogenous and typically contains a certain degree of porosity, and, in the case of sprayed metals, the coating will contain oxides of the metal. Feedstock material may be any substance that can be melted, including metals, metallic compounds, cements, oxides, glasses, and polymers. Feedstock materials can be sprayed as powders, wires, or rods. The bond between the substrate and the coating may be mechanical, chemical, or metallurgical or a combination of these. The properties of the applied coating are dependent on the feedstock material, the thermal spray process and application parameters, and posttreatment of the applied coating.

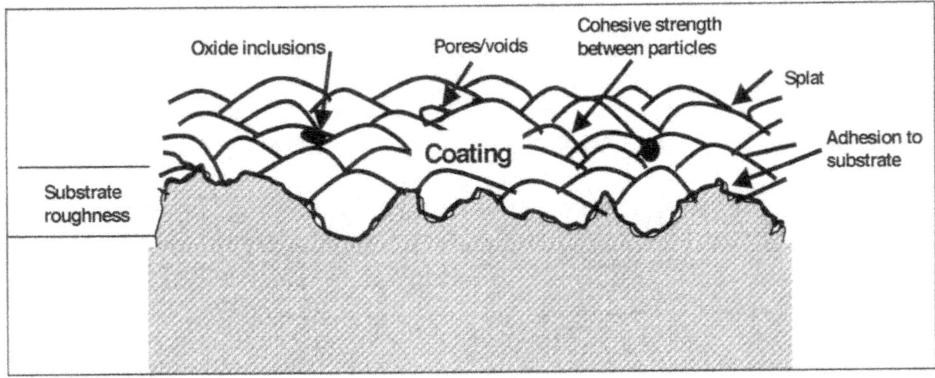

Figure 2-1 Typical cross section of a thermal spray coating

2-3. Characteristics of Thermal Spray Coatings

a. Hardness, density, and porosity. Thermal spray coatings are often used because of their high degree of hardness relative to paint coatings. Their hardness and erosion resistance make them especially valuable in high-wear applications. The hardness and density of thermal spray coatings are typically lower than for the feedstock material from which the coatings were formed. In the case of thermal spray metallic coatings, the hardness and density of the coating depend on the thermal spray material, type of thermal spray equipment, and the spray parameters. In general, the higher the particle velocity, the harder and denser the coating. Particle velocities for different thermal spray processes in descending order are detonation, high-velocity oxygen flame (HVOF), arc plasma, arc wire, and flame spray. Hardness and density may also depend on particle temperature and the type of atomization gas used. The porosity of the coating depends also on the thermal spray process, application parameters, and thermal spray material.

b. Corrosion resistance. Metallic thermal spray coatings may be either anodic or cathodic to the underlying metal substrate. Because corrosion occurs at the anode, anodic coatings will corrode in corrosive environments and the cathode will not. Anticorrosive coating systems are generally designed such that the coating material is anodic to the substrate metal. Anodic coatings will corrode or sacrifice to protect the substrate. In some cases, the corrosion resistance of the thermal spray material itself is important. For very high temperature applications and for chemical exposures, the thermal spray coating must be very corrosion resistant. For such applications, the coating provides a corrosion resistant barrier to protect the substrate. For a more complete discussion of corrosion theory please refer to Chapter 2 of EM 1110-2-3400.

c. Adhesion. Thermal spray coatings may have very high adhesion. Special coatings, used for wear resistance, that are applied by thermal spray processes with very high particle velocity can have tensile adhesions of greater than 34,000 kPa (5000 psi) as measured by ASTM C633 "Standard Test Method for Adhesion or Cohesive Strength of Flame-Sprayed Coatings." Most coatings used for infrastructure applications have adhesion values comparable to paint coatings. Typical field- and shop-applied zinc, aluminum, and zinc-aluminum alloy coatings will have adhesion ranging from 5440 to 13,600 kPa (800 to 2000 psi) as measured by ASTM D4541 "Standard Test Method for Pull-Off Strength of Coatings Using Portable Adhesion Testers."

2-4. Types of Thermal Spray Coatings

a. Corrosion resistant zinc, aluminum, and zinc-aluminum alloy coatings. Zinc, aluminum, and zinc-aluminum alloy coatings are important anticorrosive coatings because they are anodic to steel. In other words, they corrode preferentially to steel, acting as sacrificial coatings preventing the corrosion of the underlying steel substrate. Zinc is a much more active metal in this respect than aluminum. On the other hand, aluminum coatings are harder, have better adhesion, and form a protective oxide layer that prevents self-corrosion. Alloys of the two metals have properties somewhere in between, depending on the ratio of zinc to aluminum. An 85-15 (percent by weight) alloy of zinc and aluminum is a widely used thermal spray coating material because it is thought to have the best combination of attributes from both metals.

b. Polymer coatings. Thermal spray polymer or plastic coatings have been developed for infrastructure applications. Thermal spray polymers are thermoplastic powders applied by flame or plasma spray. The polymer must have a melt temperature that is conducive to thermal spray. In addition, the polymer must not polymerize, degrade, or char in the flame. Thermal spray plastics do not contain volatile organic compounds and thus are compliant for use in areas with air pollution regulations. Thermal spray polymer coatings have been used to coat steel under very cold atmospheric conditions when painting was not practical. Research has been conducted on the use of recycled plastics for polymer flame spray, and these products show some potential. There appears to be a growing interest in polymer flame spray for infrastructure applications. The Society for Protective Coatings is developing a specification for polymer flame spray, and several vendors offer equipment and polymer feedstocks.

c. Other thermal spray coatings. Other thermal spray coating materials are used for special applications. Special metal alloy coatings are commonly used for hardfacing items such as wear surfaces of farm equipment, jet engine components, and machine tools. Ferrous metal alloys are often used for restoration or redimensioning of worn equipment. Special ferrous alloys are sometimes used for high-temperature corrosion resistance. Inert ceramic coatings have been used on medical prosthetic devices and implants such as joint replacements. Conductive metal coatings are used for shielding sensitive electronic equipment against electric and magnetic fields. Ceramic coatings have also been used to produce very low-friction surfaces on near net shape components. These and other applications make thermal spray coatings a diverse industry.

2-5. Thermal Spray Processes

Thermal spray processes may be categorized as either combustion or electric processes. Combustion processes include flame spraying, HVOC spraying, and detonation flame spraying. Electric processes include arc spraying and plasma spraying.

a. Combustion processes.

(1) Flame spraying. The oldest form of thermal spray, flame spraying, may be used to apply a wide variety of feedstock materials including metal wires, ceramic rods, and metallic and nonmetallic powders. In flame spraying, the feedstock material is fed continuously to the tip of the spray gun where it is melted in a fuel gas flame and propelled to the substrate in a stream of atomizing gas. Common fuel gases are acetylene, propane, and methyl acetylene-propadiene. Air is typically used as the atomization gas. Oxyacetylene flames are used extensively for wire flame spraying because of the degree of control and the high temperatures offered by these gases. By gauging its appearance, the flame can be easily adjusted to be an oxidizing, neutral, or reducing flame. The lower temperature propane flame can be used for lower melting metals such as aluminum and zinc as well as polymer feedstocks. The basic components of a flame spray system include the flame spray gun, feedstock material and feeding mechanism, oxygen and fuel gases with flowmeters and pressure regulators, and an air compressor and regulator.

(a) Wire flame spraying. Wire flame spray is the flame process of greatest interest to the Corps of Engineers. CEGS-09971 allows for the application of aluminum, zinc, and zinc/aluminum alloy coatings using the flame spray method. Figure 2-2 shows a schematic of a typical flame spray system. Figure 2-3 depicts a typical wire flame spray gun. The wire flame spray gun consists of a drive unit with a motor and drive rollers for feeding the wire and a gas head with valves, gas nozzle, and air cap that control the flame and atomization air. Compared with arc spraying, wire flame spraying is generally slower and more costly because of the relatively high cost of the oxygen-fuel gas mixture compared with the cost of electricity. However, flame spraying systems, at only one-third to one-half the cost of wire arc spray systems, are significantly cheaper. Flame spray systems are field portable and may be used to apply quality metal coatings for corrosion protection.

(b) Powder flame spraying. Powder flame operates in much the same way as wire flame spray except that a powder feedstock material is used rather than wire and there is no atomizing air stream. The melted coating material is atomized and propelled to the surface in the stream of burning fuel gas. The powder is stored in either a gravity type hopper attached to the top of a spray gun or a larger air or inert gas entrainment type detached hopper. Powder flame spray guns are lighter and smaller than other types of thermal spray guns. Production rates for powder flame spray are generally less than for wire flame spray or arc spray. Particle velocities are lower for flame spray, and the applied coatings are generally less dense and not as adherent as those applied by other thermal spray methods. USACE use of powder flame spray should be limited to repair of small areas of previously applied thermal spray coatings and galvanizing. Figure 2-4 illustrates a typical combustion powder gun installation, and Figure 2-5 shows a powder gun cross section.

(2) HVOF spraying. One of the newest methods of thermal spray, HVOF, utilizes oxygen and a fuel gas at high pressure. Typical fuel gases are propane, propylene, and hydrogen. The burning gas mixture is accelerated to supersonic speeds, and a powdered feedstock is injected into the flame. The process minimizes thermal input and maximizes particle kinetic energy to produce coatings that are very

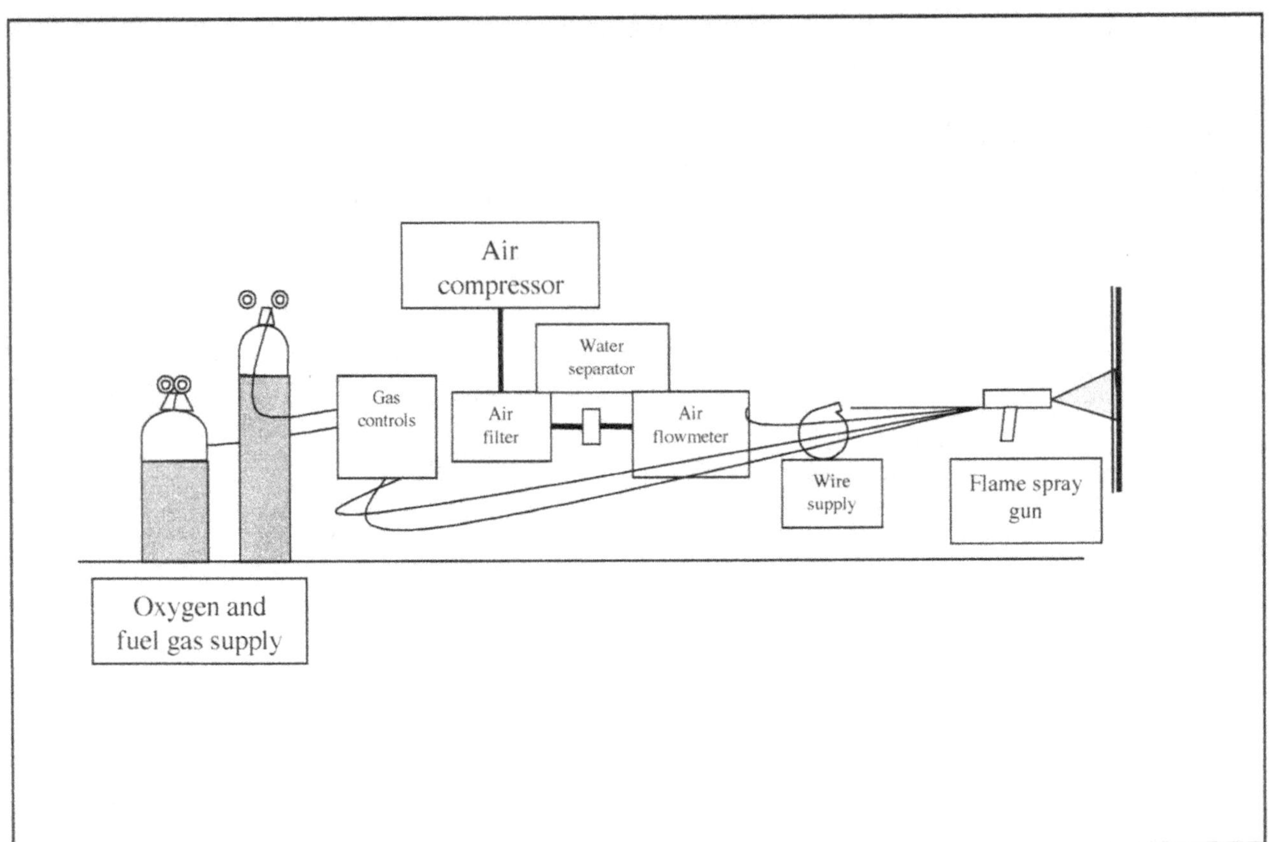

Figure 2-2. Typical flame spray system

dense, with low porosity and high bond strength. HVOF systems are field portable but are primarily used in fabrication shops. HVOF has been used extensively to apply wear resistant coatings for applications such as jet engine components. The Corps has conducted an experimental evaluation of HVOF-applied metal alloy coatings for protection against cavitation wear in hydraulic turbines.

(3) Detonation flame spraying. In detonation flame spraying, a mixture of oxygen, acetylene, and powdered feedstock material are detonated by sparks in a gun chamber several times per second. The coating material is deposited at very high velocities to produce very dense coatings. Typical applications include wear resistant ceramic coatings for high-temperature use. Detonation flame spraying can only be performed in a fabrication shop. Detonation flame spraying is not applicable for USACE projects.

b. Electric processes.

(1) Arc spraying. Arc spraying is generally the most economical thermal spray method for applying corrosion resistant metal coatings, including zinc, aluminum, and their alloys as described in CEGS-09971. Energy costs are lower and production rates are higher than they are with competing methods such as wire flame spray. Arc spraying may be used to apply electrically conductive materials including metals, alloys, and metal-metal oxide mixtures. In arc spraying, an arc between two wires is used to melt the coating material. Compressed gas, usually air, is used to atomize and propel the molten material to the substrate. The two wires are continuously fed to the gun at a uniform speed. A low voltage (18 to 40 volts) direct current (DC) power supply is used, with one wire serving as the cathode and the other as the anode. Figure 2-6 shows a typical arc spray system comprised of a DC power supply, insulated power

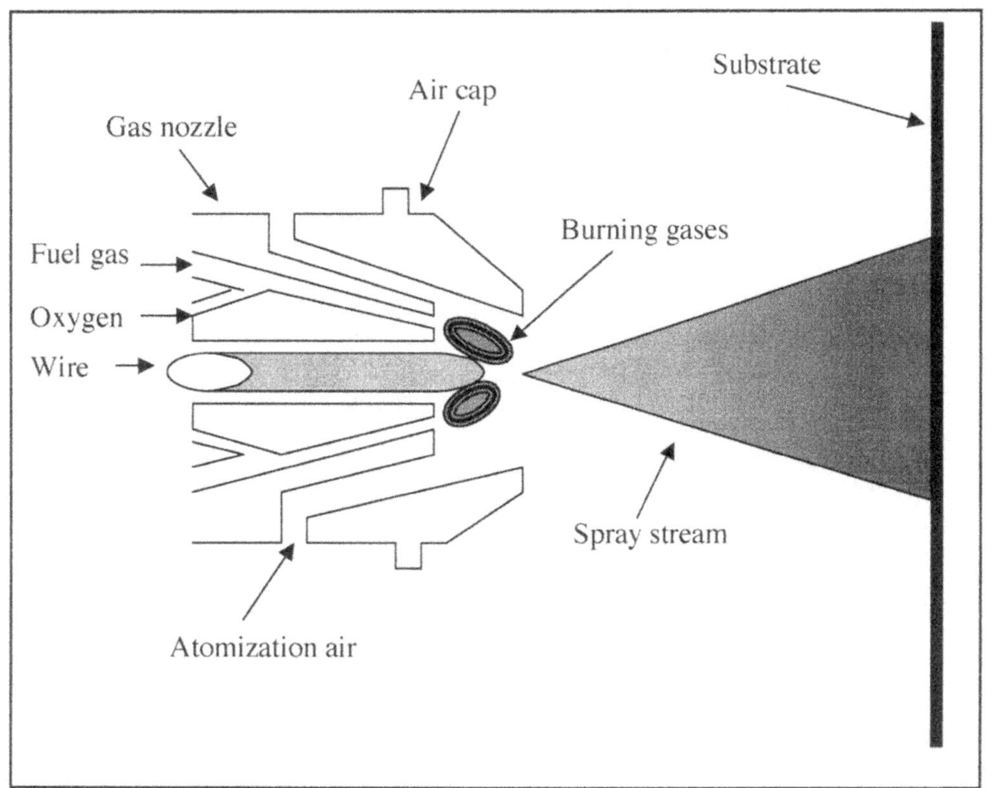

Figure 2-3. Typical flame spray gun

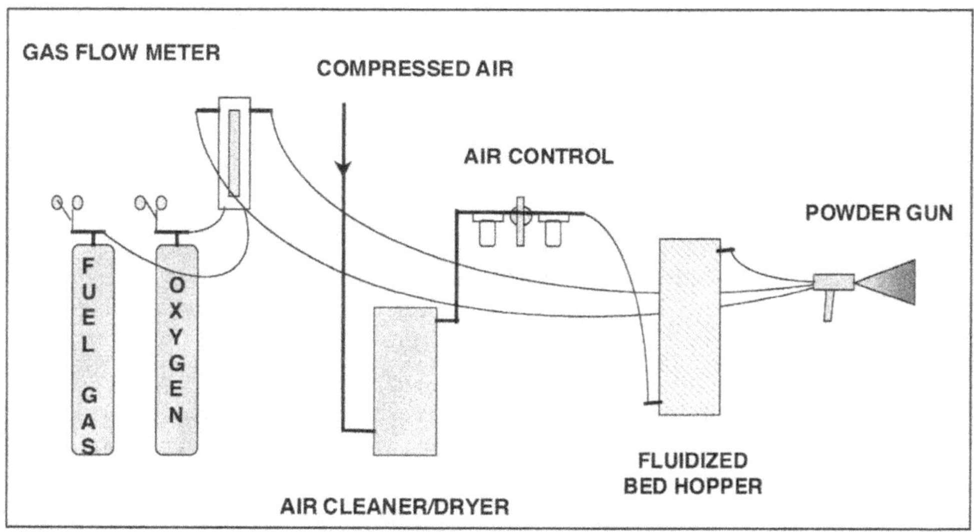

Figure 2-4. Typical combustion powder gun installation

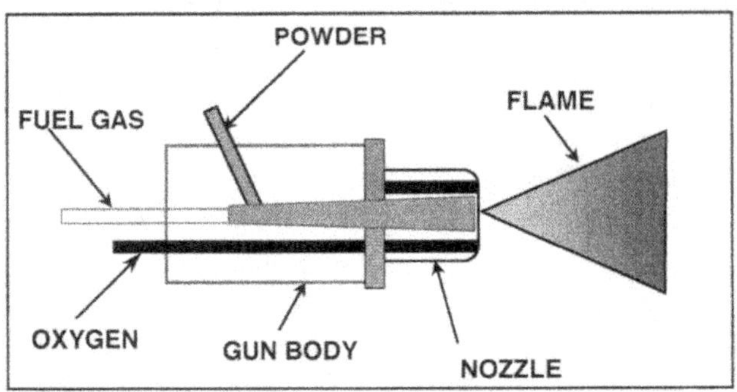

Figure 2-5. Powder gun cross section

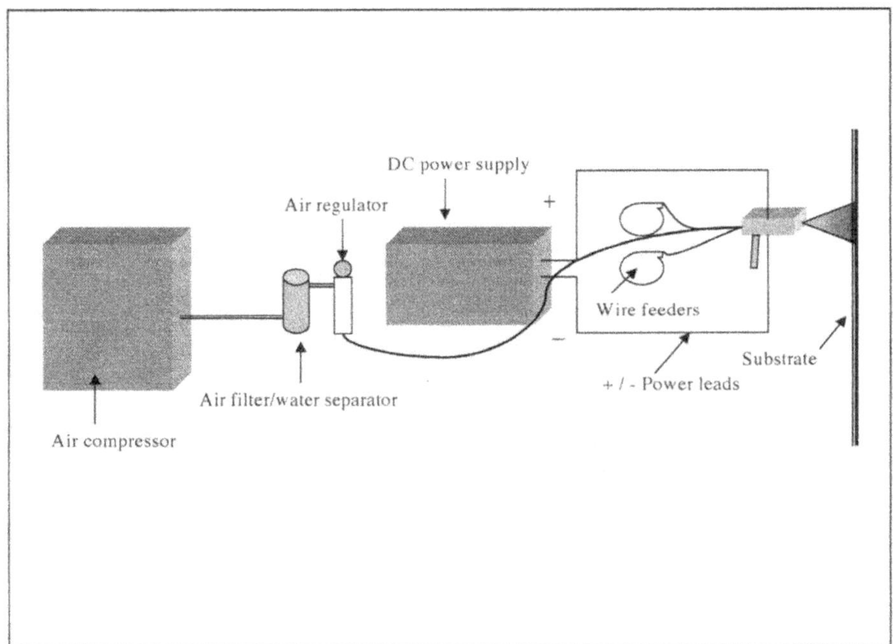

Figure 2-6. Typical two-wire arc spray system

cables, a wire feed system, a compressed-air supply, controls, and an arc spray gun. Figure 2-7 shows the components of a typical arc spray gun, including wire guides, gun housing, and gas nozzle. Coating quality and properties can be controlled by varying the atomization pressure, air nozzle shape, power, wire feed rate, traverse speed, and standoff distance. Arc sprayed coatings exhibit excellent adhesive and cohesive strength.

(2) Plasma spraying. Plasma spraying is used to apply surfacing materials that melt at very high temperatures. An arc is formed between an electrode and the spray nozzle, which acts as the second electrode. A pressurized inert gas is passed between the electrodes where it is heated to very high temperatures to form a plasma gas. Powdered feedstock material is then introduced into the heated gas where it melts and is propelled to the substrate at a high velocity. A plasma spray system consists of a power supply, gas source, gun, and powder feeding mechanism. Plasma spraying is primarily performed in fabrication shops. The process may be used to apply thermal barrier materials, such as zirconia and alumina, and wear resistant coatings such as chromium oxide.

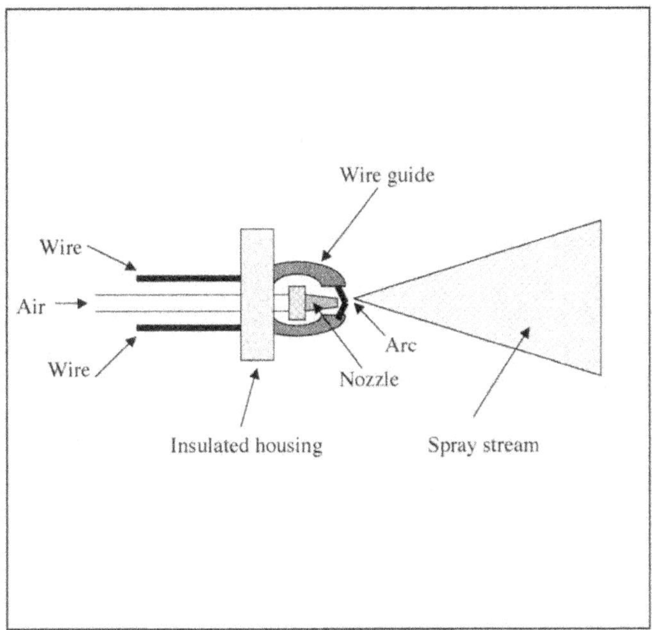

Figure 2-7. Typical two-wire arc spray gun

2-6. Thermal Spray Uses

a. Thermal spray is used for a wide variety of applications. The primary use of thermal spray coatings by the Corps is for corrosion protection. Coatings of zinc, aluminum, and their alloys are anodic to steel and iron and will prevent corrosion in a variety of service environments including atmospheric, salt- and freshwater immersion, and high-temperature applications. Coatings of aluminum are frequently used in marine environments. The U.S. Navy uses aluminum coatings for corrosion protection of many ship components. Because these materials are anodic to steel, their porosity does not impair their ability to protect the ferrous metal substrate. Zinc and zinc-aluminum alloy coatings may corrode at an accelerated rate in severe industrial atmospheres or in chemical environments where the pH is either low or high. For this reason these materials are typically sealed and painted to improve their performance.

b. Cathodic coatings such as copper-nickel alloys and stainless steels can also be used to protect mild steel from corrosion. These materials must be sealed to prevent moisture migration through the coating. These metals are particularly hard and are often used for applications requiring both corrosion and wear resistance.

c. Aluminum coatings are often used for corrosion protection at temperatures as high as 660 °C (1220 °F).

d. Thermal spray deposits containing zinc and/or copper can be used to prevent both marine and freshwater fouling. Zinc and 85-15 zinc-aluminum alloy coatings have been shown to prevent the significant attachment and fouling by zebra mussels on steel substrates. Because these coatings are long lived and prevent corrosion, their use is recommended for Corps structures. Copper and brass coatings have also been shown to be effective antifoulants but should not be used on steel due to the galvanic reaction between the two.

e. Zinc thermal spray coatings are sometimes used to prevent the corrosion of reinforcing steel imbedded in concrete. For such applications, the zinc is deposited onto the concrete and is electrically connected to the steel.

f. Thermal spray coatings are frequently used to repair surfaces subject to wear. A common application is the redimensioning of rotating shafts. Metal is sprayed onto the part as it is rotated on a lathe. The rebuilt part can then be machined to the required diameter. Similarly, thermal spray deposits can be used to recontour foundry molds or to repair holes.

g. Thermal spray coatings are also used for electrical applications. Conductive metals such as copper can be used for conductors. Ceramic materials may be used for electrical insulation. Conductive metals are also used to magnetically shield sensitive electronics.

h. Very hard and dense thermal spray deposits have been used on an experimental basis as cavitation resistant materials and in conjunction with weld overlays as a repair technique.

Chapter 3
Thermal Spray Materials

3-1. Introduction

This chapter is intended to provide the engineer with an understanding of how thermal spray coatings are specified, procured, and tested prior to being applied. Thermal spray coating materials can be specified by describing the composition of the wire or powder, by product name and manufacturer, and by citing the material descriptions herein. Thermal spray wire and powder testing, including sampling procedures, material identification, and coating performance testing, is critical in establishing whether the supplied materials meet the composition requirements and whether they will provide the desired level of corrosion protection to the structure. This chapter will provide the engineer with an understanding of the various tests that can be performed and what the test data mean in terms of thermal spray coatings performance.

3-2. Specifications

Thermal spray coating materials can be specified by product name and manufacturer or by using a material description. Each method of specification will be discussed.

a. Specification by product name/manufacturer. The product name of a manufacturer is one way to specify a coating material. Private industry often specifies thermal spray materials by product name/manufacturer; however, the USACE does not purchase materials in this way. Specifying thermal spray materials by product name/ manufacturer can be beneficial when a specific thermal spray coating material has proven successful. Technical information and advice on applying the coating material are typically available from the manufacturer. Specification by product name/manufacturer limits competition and may result in higher material costs. Refer to CFR 48 1-10.002 and ER 1110-2-1200 concerning restrictions on specifying proprietary products.

b. Specification by material description. A material description that provides the compositional, mechanical, and physical characteristics of the thermal spray material may be used. This method should generally be used to specify thermal spray materials for USACE projects. In addition to thermal spray material compositional, mechanical, and physical characteristics, the description also provides a means for material classification, acceptance, certification, testing, manufacture, wire sizes, packaging forms, feedstock identification, and marking of packages.

3-3. Procurement

Thermal spray feedstock materials are typically purchased by the contractor, and, in such cases, it is the contractor's responsibility to procure material that meets the specification requirements. The USACE does not generally provide thermal spray materials to a contractor because the USACE would be responsible for storage, short or excess supply, timely delivery, and waste disposal.

3-4. Classification

Thermal spray materials are classified based on chemical composition and mechanical and physical characteristics. Figure 3-1 shows the nomenclature used to designate the thermal spray wire and ceramic rod

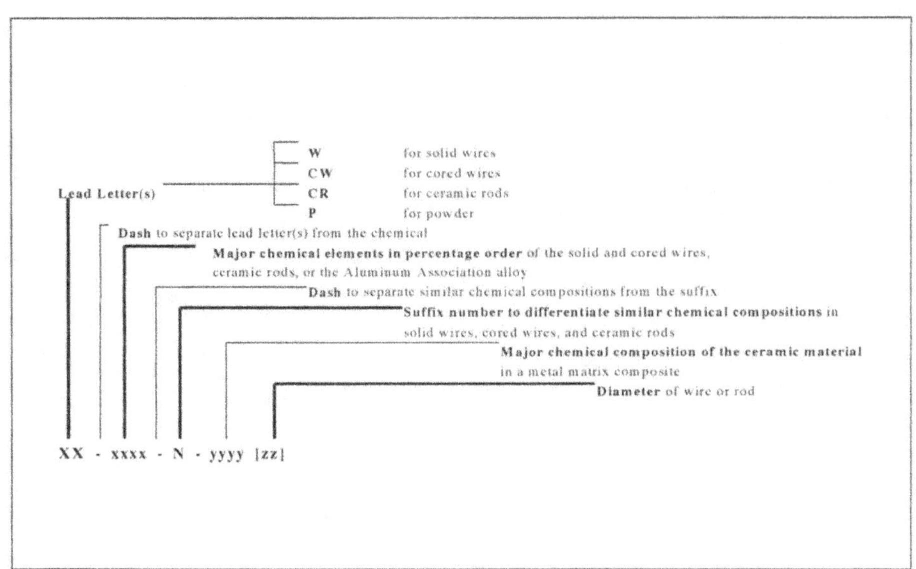

Figure 3-1. Thermal spray feedstock designation nomenclature

feedstock. The description does not address powder feedstock materials. Powder feedstocks are not often used for large-scale production activities. However, if powder feedstock is to be used, it is recommended that the material be held to the same compositional requirements as the equivalent wire material.

3-5. Acceptance

Criteria for acceptance, quality control, and level of testing are described by ANSI/AWS A5.01, "Filler Metal Procurement Guidelines." Acceptance of thermal spray materials is based on the fulfillment of the testing requirements described by ANSI/AWS A5.01. A level of testing as defined in ANSI/AWS A5.01 is ordinarily specified in the procurement document. If the level of testing is not specified, then the manufacturer's standard testing level is assumed. This level of testing is designated as Schedule F in Table 1 of ANSI/AWS A5.01. In general, for USACE projects, Schedule H level of testing from ANSI/AWS A5.01 should be used as the basis for accepting thermal spray feedstock materials. Schedule H level of testing is chemical analysis only, for each lot of material supplied.

3-6. Certification

The manufacturer should certify that the thermal spray material meets the requirements of the material description. Certification implies that the required testing was performed on material representative of that being shipped and that the product conforms to the testing requirements of the specification. Representative material is defined as any material from any production run of the same class of material with the same formula. Certification does not necessarily mean that tests were run on the actual material being supplied.

3-7. Sizes

Thermal spray wire specified for USACE projects will generally be supplied in 3.2-mm (1/8-in.) and 4.8-mm (3/16-in.) diameters. USACE contractors should be allowed to purchase wire sizes appropriate for the equipment to be used on the job.

3-8. Packaging

Thermal spray wire is supplied in coils with and without support, spools, and drums. Standard package weights and dimensions are common. Nonstandard sizes and weights may be supplied as agreed between the supplier and purchaser. The dimension and weight of coils without support are by agreement between the purchaser and supplier. In general, for USACE jobs, the contractor should be allowed to procure wire in standard packages consistent with the requirements of the work to be performed. Appropriate packaging is designed to protect the thermal spray material from damage during shipment and storage.

3-9. Identification and Marking

All thermal spray materials should be properly identified by marking each spool, coil, or drum. Coils without support are marked by an attached tag. Coils with support are marked on the support itself. Spools are marked on the flange of the spool. Drums are marked on the top and side. As a minimum, markings generally contain information about the product, including the material classification, manufacturer's name and product designation, size and weight, heat number, and precautionary information.

3-10. Manufacture

Thermal spray wire may be manufactured by any process, provided that the material meets the requirements of the specification. The manufactured wire should have a smooth finish, free of defects that may affect the feeding of the wire to the thermal spray gun. Such defects include slivers, depressions, scratches, scale, laps, and surface contaminants. A small amount of lubricant may be used on some wire feedstocks to improve wire feeding. Wire may be welded together as supplied to provide for a single continuous wire in a package. Welded wire should be smooth and should not interfere with feeding. The temper of thermal spray wire should allow for continuous and smooth feeding of the wire during spray application. The wire should be wound such that there are no kinks, waves, sharp bends, or overlaps that interfere with wire feed in the thermal spray equipment. The free end of the wire should be secured to prevent unwinding and should be marked for easy identification.

3-11. Testing

Thermal spray testing generally falls into two categories, testing of the feedstock materials and testing of the applied coating. This section addresses the testing of feedstock materials only.

a. Chemical composition. Table 3-1 gives the chemical composition requirements for aluminum, zinc, and alloy thermal spray wires. Chemical composition is determined by various ASTM test methods utilizing emission spectrochemical analysis, inductively coupled plasma spectroscopy, and wet chemistry techniques.

b. Surface appearance. Surface appearance is determined by visual examination of the wire for defects that may interfere with the smooth feeding and application of the wire. Such defects are described above in paragraph 3-10.

c. Cast and helix. This test involves cutting, from a standard package, a specimen that is long enough to form a single loop. The loop of wire, when placed on a flat surface, should form a circle with a diameter of at least 38 cm (15 in.). The loop should not rise more than 2.5 cm (1 in.) above the flat surface. Cast and helix requirements are not applicable to soft alloy wires, including aluminum, copper, and zinc. Most of the feedstock materials used by USACE are soft alloys, and the cast and helix test is not generally a problem.

Table 3-1
Chemical Composition Classification of Aluminum, Zinc, and Alloy Thermal Spray Wires

		Composition, Wt % (a)										Other Elements	
Classification	Common Name	Al	Cr	Cu	Fe	Mn	Pb	Si	Sn	Ti	Zn	Element	Amount
W-Al-1100	1100 Aluminum	99.00 min	...	0.05-0.20	0.95 (Fe+Si)	0.05	...	0.95 (Fe+Si)	...	--	0.1	.(b), (c)..	(b), (c)...
W-Al-1350	1350 Aluminum	99.50 min	0.01	0.05	0.40	0.01	...	0.10	...	0.02 (V+Ti)	0.05	Ga	0.03
W-Al-4043	4043 Silicon Aluminum	rem	--	0.30	0.8	0.05	...	4.5-6.0	...	0.20	0.10	Mg (b)	0.05
W-Al-4047	4047 Silicon Aluminum	rem	...	0.30	0.8	0.15	...	11.0-13.0	...	...	0.20	Mg (b)	0.10
W-Al-5356	5356 Mg Aluminum	rem	0.05-0.20	0.10	0.40	0.05-0.20	...	0.25	...	0.06-0.20	0.10	Mg (b)	4.5-5.5
W-Al-Al_2O_3	Al MMC (d)	88 min	...	...	...	...	...	...	...	...	...	Al_2O_3 (b)	8-12
W-Zn-1	99.99 Zinc	0.002	...	0.005	0.003	...	0.003	...	0.001	...	99.99 min	Cd	0.003
W-Zn-2	99.9 Zinc	0.01	...	0.02	0.02	...	0.03	...	...	...	99.9 min	Cd	0.02
W-ZnAl-1	98/2 Zinc-Aluminum	1.5-2.5	...	...	...	...	...	...	...	...	rem	non Zn/Al	0.1
W-ZnAl-2	85/15 Zinc-Aluminum	14.0-16.0									Rem	Other	0.05

Notes: (a) Single values shown are maximum percentages unless a minimum is specified.
(b) 0.0008 Be maxi.
(c) Others: 0.05 max each, 0.15 max total.
(d) Vol% Aluminum Assn. 1060 Alloy (99.6% pur Al) with addition of 8-12 vol% Al_2O_3 powder, 8-10 micron diameter.

Chapter 4
Thermal Spray Coating Cost and Service Life

4-1. Introduction

This chapter contrasts paint and thermal spray coatings based on cost and expected service life. Both paint and thermal spray coatings may be used to provide corrosion protection for most civil works applications. The use of thermal spray coatings is preferred on the basis of fitness-for-purpose for a few specific applications, including corrosion protection in very turbulent ice- and debris-laden water, high-temperature applications, and zebra mussel resistance. Thermal spray coatings may also be selected because of restrictive air pollution regulations that do not allow the use of some paint coatings specified in CEGS-09965. For all other applications, the choice between thermal spray and paint should be based on life-cycle cost.

4-2. Cost

a. Whenever possible, coating selection should be based on life-cycle cost. In reality, the engineer must balance competing needs and may not always be able to specify the least expensive coating on a life-cycle cost basis. Because of their somewhat higher first cost, thermal spray coatings are often overlooked. To calculate life-cycle costs, the installed cost of the coating system and its expected service life must be known. Life- cycle costs for coating systems are readily compared by calculating the average equivalent annual cost (AEAC) for each system under consideration.

b. The basic installed cost of a thermal spray coating system is calculated by adding the costs for surface preparation, materials, consumables, and thermal spray application. The cost of surface preparation is well known. The cost of time, materials, and consumables may be calculated using the following stepwise procedure:

(1) Calculate the surface area (*SA*). (SA = length × width)

(2) Calculate the volume (V) of coating material needed to coat the area. ($V = SA \times$ coating thickness)

(3) Calculate the weight of the material to be deposited (Wd). The density (D) of the applied coating is less than that of the feedstock material. A good assumption is that the applied coating is about 90 percent of the density of the feedstock material. The densities of aluminum, zinc, and 85-15 zinc-aluminum wire are 2.61 g/cm^3 (0.092 lb/in.^{3}), 7.32 g/cm^3 (0.258 lb/in.^{3}), and 5.87 g/cm^3 (0.207 lb/in.^{3}), respectively. ($Wd = V \times 0.9\ D$)

(4) Calculate the weight (W) of material used. Estimates of deposition efficiency (DE) for various materials and thermal spray processes are given in Chapter 7, Table 7-2. ($W = Wd/DE$)

(5) Calculate the spray time (T). Spray rates (SR) for various materials and thermal spray processes are given in Chapter 7, Table 7-4. ($T = W/SR$)

(6) Calculate electricity or oxygen and fuel gas consumption (C). Typical consumption rates (CR) for electricity, fuel gas, and oxygen are available from equipment manufacturers. ($C = CR \times T$)

(7) Calculate cost of materials (CM). ($CM = W \times$ cost per unit weight)

(8) Calculate cost of application (*CA*). ($CA = T \times$ unit labor cost)

(9) Calculate cost of consumables (*CC*). ($CC = T \times$ unit cost of consumable)

(10) Calculate total cost (*TC*) of thermal spray coating. ($TC = CM + CA + CC$)

c. Other factors that increase the cost of thermal spray and other coating jobs include the costs of containment, inspection, rigging, mobilization, waste storage, and worker health and safety.

d. The Federal Highway Administration (FHWA) (1997) compared the performance of a number of coating systems, including paints and thermal spray. Coating life expectancies were estimated based on performance in an aggressive marine atmospheric exposure and a mildly corrosive environment. Installed and life-cycle costs were calculated for each coating system for each exposure. Average equivalent annual costs were calculated based on a 60-year structure life. For the more severe marine atmospheric exposure, thermal spray coatings of aluminum, zinc, and 85-15 zinc-aluminum alloy were the most cost-effective coatings. For the less severe mildly corrosive atmospheric exposure, thermal spray was no more or less cost effective than other coating options.

4-3. Service Life

There are many documented examples of thermal spray coatings of zinc and aluminum with very long service lives. Service life depends on thermal spray coating thickness and the exposure environment. There does not appear to be a significant difference in the long-term performance of thermal spray coatings applied by different processes, including arc, wire flame, and powder flame spray. Thermal spray zinc coatings applied at thicknesses of 250 µm (0.10 in.) have performed for more than 40 years in atmospheric exposures. Zinc thermal spray coatings in potable water tanks have lasted longer than 30 years. The FHWA (1997) report estimates a service life of 30 and 60 years for 85-15 zinc-aluminum alloy coating (150 µm (0.006 in.)) in severe marine and mildly corrosive atmospheres, respectively. USACE has experience with 85-15 zinc-aluminum alloy coatings (400 µm (0.016 in.)) providing 10 years of service in very turbulent ice- and debris-laden water. Table 4-1 provides typical service lives of paint coatings and predicted service life of thermal spray coatings for selected USACE applications. The tabulated service lives are given as the time to first maintenance.

Table 4-1
Predicted Service Life for Selected Thermal Spray Applications

Application	Paint System[a]	Typical Paint Service Life	Thermal Spray System Number[b]	Predicted Thermal Spray Service Life, years
Penstocks	Coal tar epoxy	20 – 30	6-Z-A	30 – 40
Tainter gates	Vinyl zinc-rich	20 – 25	6-Z-A	25 – 35
Tainter and roller gates (interior)	Vinyl	25 – 40	6-Z-A	40 – 50
Tainter gates (very turbulent ice- and debris-laden water)	Vinyl zinc-rich	1 – 2	6-Z-A	8 – 12
Roller gates	Vinyl zinc-rich	25 – 30	6-Z-A	30 – 40
Service bridges	Alkyd/phenolic	10 – 15	5-Z-A or 2-Z	50 - 60
Sector gates (seawater)	Epoxy zinc-rich/coal tar epoxy	15 – 20	8-A	20 - 40

[a] Paint systems described in CEGS-09965:
Coal tar epoxy = Paint System No. 6
Vinyl zinc-rich = Paint System Nos. 3-A-Z, 5-C-Z, and 5-E-Z
Vinyl = Paint System No. 4
Alkyd/phenolic = Paint System No. 2
Epoxy zinc rich/coal tar epoxy = Paint System No. 6-A-Z

[b] 6-Z-A = 400 μm (0.016 in.) 85-15 zinc-aluminum alloy
5-Z-A = 300 μm (0.012 in.) 85-15 zinc-aluminum alloy
2-Z = 300 μm (0.012 in.) zinc
8-A = 250 μm (0.010 in.) aluminum

Chapter 5
Thermal Spray Coating Selection

5-1. Introduction

A systematic approach to coating selection for new construction and maintenance thermal spraying is described in this chapter. Paragraph 5-2 discusses criteria important to the selection of thermal spray coatings including the service environment, expected longevity, ease of application, and maintainability. Paragraph 5-3 discusses the relative merits of paint coatings and thermal spray coatings including durability and environmental considerations. Subsequent paragraphs discuss thermal spray coating systems for specific USACE applications.

5-2. Service Environments

Foreknowledge of the environmental stresses to which the protective coating system will be exposed is critical for proper selection of the coating system. This is true of both paint and thermal spray coating systems. Exposure environments typically encompass one or more of the following environmental stresses: extremes of temperature, high humidity, immersion, extremes of pH, solvent exposure, wet/dry cycling, thermal cycling, ultraviolet exposure, impact and abrasion, cavitation/erosion, and special exposures. The service environment is the single most important consideration in the selection of a coating system.

a. Extremes of temperature. Most exposure environments show some variability in temperature. Normal atmospheric service temperatures in northern latitudes of the continental United States vary from -23 to 38 °C (-10 to 100 °F). Temperatures for immersion exposure show somewhat less variation and typically range from about -1 to 27 °C (30 to 80 °F). These normal variations in temperature are relatively insignificant to the performance of thermal spray coatings of zinc, aluminum, and their alloys. Paint coating performance is generally more sensitive to these normal extremes of temperature. Some components, such as the stacks of floating plants, may be subject to higher than normal atmospheric temperatures. With some exceptions, most paint coatings will not perform well at these elevated service temperatures. Most alkyd paints such as CID A-A-2962 will tolerate temperatures up to only about 120 °C (250 °F). Special black bituminous coatings such as CID A-A-3054 will withstand temperatures up to 204 °C (400 °F). Color pigmented modified and unmodified silicone coatings may °F) and 315 °C (600 °F), respectively. Aluminum and carbon black pigmented silicone coatings may perform at temperatures as high as 650 °C (1200 °F). Special ceramic frit coatings may perform at temperatures of 760 °C (1400 °F). SSPC Paint 20 Type I-B or I-C inorganic zinc-rich coatings can usually perform at temperatures up to 400 °C (750 °F). Thermal spray coatings of aluminum, zinc, and their alloys will provide long-term performance superior to paint coatings at temperatures approaching their respective melting points of 660 °C (1220 °F) and 420 °C (788 °F). Because of its excellent temperature resistance and corrosion protection, the aluminum thermal spray system 8-A from CEGS-09971 is recommended for applications where temperatures will exceed 400 °C (750 °F). Below this temperature, the specifier may elect to use paint system number 10 from CEGS-09965 which consists of two coats of SSPC Paint 20 Types I-B or I-C.

b. High humidity. High humidity is often accompanied by condensation, which is considered to approximate the severity of freshwater immersion. All of the thermal spray systems described in CEGS-09971 will also perform well in high-humidity condensate exposures. System 5-Z-A is the recommended thermal spray system for high-humidity condensate environments. Typically, high-performance paint systems such as the epoxy and vinyl systems described in CEGS-09965 are specified for high-humidity applications. Because paint systems are generally less costly to apply, they are more likely to be used for these types of exposures. However, thermal spray system 5-Z-A should have a longer service life than paint coatings for this application.

c. Immersion. Immersion exposures range from immersion in deionized water to immersion in natural waters, including fresh water and seawater. Ionic content and pH contribute to the corrosivity of immersion environments. Typical sealers and topcoats are vinyl paints V-766e, V-102e, V-103c, and V-106d and coal tar epoxy coating C-200A. Several of the epoxy systems and all of the vinyl systems described in CEGS-09965 are appropriate for various immersion exposures depending on whether the water is fresh or salt and the degree of impact and abrasion. The epoxy systems are preferred for saltwater exposures, while the vinyl systems are generally preferred for freshwater exposures, especially where the level of impact and abrasion is significant.

(1) Seawater. Aluminum thermal spray system 8-A described in CEGS-09971 is recommended for seawater immersion. Aluminum thermal spray has been used extensively by the offshore oil industry to protect immersed and splash zone platform components from corrosion. Aluminum thermal spray is thought to perform better in seawater immersion without an organic sealer and paint topcoat.

(2) Fresh water. Thermal spray systems 5-Z-A and 6-Z-A are recommended for freshwater immersion, with 6-Z-A being the preferred choice for more severe exposures. These systems can be used either with or without sealers and topcoats.

d. Extremes of pH. Extremes of pH, such as strongly acidic or alkaline environments can greatly affect coating performance. The coating must be relatively impermeable to prevent migration of the acidic or alkaline aqueous media to the substrate, and the coating material itself must be resistant to chemical attack. Thermal spray coatings of aluminum, zinc, and their alloys may perform poorly in both high and low pH environments. Both metals show increased solubility as pH increases or decreases from the neutral pH of 7. Thermal spray aluminum and zinc may be used in acidic or alkaline environments provided that they are sealed and topcoated with vinyl or epoxy coatings. Unsealed zinc thermal spray coatings are suitable for pHs of 6 to 12 and aluminum thermal spray coatings for pHs of 4 to 8.5. Thermal spray coatings containing zinc or aluminum should not be used in chemical environments where they may be exposed to strong acids such as battery acids. Alkyd paints generally have poor resistance in alkaline environments. The epoxy and vinyl systems described in CEGS-09965 perform well in mildly acidic and alkaline environments. Topcoats with aluminum pigmentation should not generally be used in these exposures. Organic coatings and linings, as well as special inorganic building materials, should be used in highly alkaline or acidic environments.

e. Solvent exposure. Solvent exposure covers a wide variety of solvent types. Thermal spray metal coatings are essentially unaffected by solvent exposure and are good candidates for service in such environments. Some owners exclude the use of all zinc-containing coatings from use in aviation fuel storage tanks because metal contamination may affect the performance of the fuel. A solvent exposure that may be harsher on thermal spray coatings than anticipated is petroleum storage tanks where a layer of corrosive water can collect on the inside bottom of the tank. The water results in a much more corrosive environment than would be assumed if only the petroleum product was present. Some blends of organic solvents or natural petroleum products may also be acidic, which may affect thermal spray coating performance. Normal exposures to organic products such as cleaning solvents, lubricants, and hydraulic fluids should not preclude the use of thermal spray coatings of aluminum, zinc, and their alloys at USACE projects. The performance of paint coatings in solvent exposures depends on the coating type and the solvent species. Specific paint types, such as epoxies, are more solvent resistant than others. Some solvent types are more aggressive than others, independent of coating type.

f. Wet/dry cycling. Alternating wet and dry conditions are normal for most atmospheric exposures and, as such, most coating systems will provide adequate protection under such conditions. Thermal spray metal coatings will provide excellent performance under normal atmospheric conditions. Sealing and topcoating of the thermal spray coating is not generally necessary for such simple exposures. Generally, coating system selection will depend more on other stresses in the environment than on simple wet/dry cycling.

g. Thermal cycling. Thermal cycling may result from normal diurnal temperature variations as well as temperature changes found in operating machinery and process vessels. Thermal cycling induces stresses within the coating. Thermal sprayed metal coatings are more apt to have coefficients of expansion similar to the substrate; therefore, their relative inflexibility does not cause them to fail under normal conditions of thermal cycling.

h. Ultraviolet exposure. Resistance to ultraviolet (UV) radiation induced degradation is an important aspect of coating performance. All thermal sprayed metallic coatings are essentially unaffected by UV radiation. Organic sealers and topcoats used over thermal spray coatings will be affected the same as any other paint material of the same type. Organic paint coatings are affected by UV radiation to varying degrees. Depending on the coating resin and pigmentation types, UV degradation may result in loss of gloss, color fading, film embrittlement, and chalking. Certain paints, including silicone and aliphatic polyurethane coatings, exhibit superior UV resistance. Some coatings, including most epoxies and alkyds, have fairly poor UV resistance.

i. Impact and abrasion. Impact and abrasion are significant environmental stresses for any coating system. Abrasion is primarily a wear-induced failure caused by contact of a solid material with the coating. Examples include foot and vehicular traffic on floor coatings, ropes attached to mooring bitts, sand suspended in water, and floating ice. When objects of significant mass and velocity move in a direction normal to the surface as opposed to parallel, as in the case of abrasion, the stress is considered impact. Abrasion damage occurs over a period of time while impact damage is typically immediate and discrete. Many coating properties are important to the resistance of impact and abrasion including good adhesion, toughness, flexibility, and hardness. Thermal spray coatings of zinc, aluminum, and their alloys are very impact resistant. Zinc metallizing has only fair abrasion resistance in immersion applications because the coating forms a weakly adherent layer of zinc oxide. This layer is readily abraded, which exposes more zinc, which in turn oxidizes and is abraded. Thermal spray coating system 6-Z-A described in CEGS-09971 is considered to be the most impact/abrasion resistant of all of the Corps' coating systems. Application of this system to tainter gates in very harsh environments has been shown to be highly effective. The vinyl paint systems described in CEGS-09965 are particularly resistant to impact damage caused by ice and floating debris, but are less resistant than metallizing. The epoxy systems are somewhat brittle and are not nearly as resistant to impact damage as are the vinyls.

j. Cavitation/Erosion. More severe than impact and abrasion environments are exposures involving cavitation and erosion. Cavitation results when very-high-pressure air bubbles implode or collapse on a surface. The pressures involved can be very high (413,000 to 1,960,000 kPa (60,000 to 285,000 psi)) and destructive. Metallic components of hydraulic equipment such as hydroelectric turbines, valves and fittings, flow meters, hydrofoils, pumps, and ship propellers are particularly susceptible to cavitation damage. Low, medium, and high severity of cavitation have been defined based on an 8000-hr operating year. Low cavitation is defined as loss of carbon steel in the range 1.6 to 3.2 mm (1/16 to 1/8 in.) over a 2-year period. Medium cavitation is loss of austenitic stainless steel at greater than 1.6 mm (1/16 in.) per year. High cavitation is loss of austenitic stainless steel greater greater than 3.2 mm (1/8 in.) in a 6-month period. The standard method of repairing cavitation damage is to remove corrosion products by gouging with an electric arc and then grinding the damaged area. The cleaned area is then filled with weld metal and redimensioned by grinding. This method is very time consuming and expensive. Very hard and dense thermal spray deposits applied by HVOF spray have been used on an experimental basis as cavitation resistant materials and in conjunction with weld overlays as a repair technique.

k. Special exposures. Special exposures may include the coating of surfaces governed by the Food and Drug Administration (FDA) and National Sanitation Foundation (NSF) for food and potable water contact, respectively. Guide specifications CEGS-09965 and CEGS-09971 do not address either of these applications. Another special exposure is the use of coatings to prevent macrofouling caused by either marine fouling organisms or zebra mussels. Many of the coatings used to control fouling contain a toxin, which must be registered with the Environmental Protection Agency under the requirements of the Federal Insecticide, Fungicide, and Rodenticide Act (FIFRA). Corps guidance documents do not address the use of coatings to control fouling organisms, however thermal spray systems containing zinc and/or copper are known to be effective zebra mussel deterrents. Thermal spray coating systems 6-Z-A and 3-Z described in CEGS-09971 are recommended control coatings for zebra mussels on immersed steel and concrete surfaces. Neither material requires FIFRA registration.

5-3. Other Considerations in Coating Selection

The specifier should also consider other aspects of the proposed coating job in order to select the most appropriate coating system. Other factors discussed below include limits on surface preparation, ease of application, regulatory requirements, field conditions, maintainability, and cost.

a. Limits on surface preparation. Coating selection may be limited by the degree or type of surface preparation that can be achieved on a particular structure or structural component. Because of physical configuration or proximity to sensitive equipment or machinery, it may not always be possible to abrasive-blast a steel substrate. In such cases, other types of surface preparation, such as hand tool or power tool cleaning, may be necessary, which, in turn, may place limits on the type of coatings that may be used. In some cases, it may be necessary to remove the old coating by means other than abrasive-blasting, such as power tools, water jetting, or chemical strippers. These surface preparation methods do not impart a surface profile that is needed by some types of coatings to perform well. In the case of thermal spray coatings, a high degree of surface preparation is essential. This kind of preparation can only be achieved by abrasive blasting with a good quality, properly sized angular blast media. Thermal spray should never be selected for jobs where it is not possible to provide the highest quality surface preparation.

b. Ease of application. Coating selection may be limited by the ability of the applicator to access the surfaces to be coated. This usually is the result of the physical configuration or design of the structure. Items of limited access such as back-to-back angles, cavities, and crevices may be difficult if not impossible to coat. Most items that can be coated by paint spray application may also be coated by thermal spray. Both methods require about the same amount of access area for hoses, maneuvering, and standoff distance. As a rule of thumb, if access to the surface allows proper blast cleaning, then thermal spray application is feasible. Thermal spray coatings perform best when sprayed in a direction normal to the surface and within a particular range of standoff distances from the substrate. Application at an angle of less than 45 deg to the vertical is not recommended. Maximum and minimum standoff distances depend on the material being applied, the manufacturer, and the type of thermal spray equipment. If the standoff distance and spray angle cannot be maintained within the specified range, hand application of a paint coating may be necessary.

c. Regulatory requirements. The use of paint coatings is regulated in terms of the type and amounts of solvents or volatile organic compounds (VOC) they contain. Certain types of solvents, such as water and acetone, are exempt from these regulations because they do not contribute to the formation of photochemical pollution or smog in the lower atmosphere. Regulations vary by geographic location and by industry. Different rules apply for architectural and industrial maintenance painting, marine painting, and miscellaneous metal parts painting. USACE field painting is considered architectural and industrial maintenance painting. Shop painting performed by a fabricator is considered miscellaneous metal parts painting. Painting of a floating plant in a shipyard or dry dock facility is considered marine painting. The specifier should consult with local and state officials to determine which rules, if any, affect the proposed coating work. There are no VOC emissions associated with the use of thermal spray coatings, and their use is not regulated by any such rule. Thermal spray coatings offer an excellent VOC compliant alternative to paint coatings for many applications. The sealers and topcoats recommended for thermal spray systems are not exempt from VOC-type regulations. The thermal spray coatings will often perform just as well without the sealers and topcoats, which can therefore be omitted for reasons of compliance with air pollution regulations. It should also be noted that there are typically low VOC paint coating alternatives for most applications. The relative merits of these products should be weighed against those of the zero VOC thermal spray coating systems.

d. Field conditions. The conditions under which the coating work will be performed are another important consideration in coating selection. Certain atmospheric conditions, including high humidity and condensation, precipitation, high winds, and extreme cold or heat, place severe limitations on any type of coating work.

(1) Moisture on the surface should always be avoided to the greatest extent possible. Certain types of paint are more tolerant of small amounts of water on the surface and should be specified for work where such conditions cannot be avoided. Thermal spray metal coatings should never be applied if moisture is present on the surface.

(2) High winds may affect the types of surface preparation and coating application methods that are practical for a given job. High winds will tend to carry surface preparation debris and paint overspray longer distances. This problem can be avoided by using methods other than open abrasive blasting and spray application of paints.

(3) The pot life of multicomponent catalyzed coatings such as epoxies can be greatly reduced by high atmospheric temperatures. High ambient air and surface temperatures can also adversely affect paint application and the subsequent performance of the coating; for example, vinyl paints are prone to dry spray at high temperatures. Most paints should not be applied below a certain minimum temperature because they will not cure or dry. Most epoxy paints should not be applied when ambient and substrate temperatures are below 10 °C (50 °F); however, there are some specialized epoxy coatings that can be applied at temperatures as low as -7 °C (20 °F). Latex coatings should never be applied when temperatures are expected to fall below 10 °C (50 °F) during application and drying. Vinyl paints can be applied at quite low temperatures compared with most paints. Vinyl application at 0 °C (32 °F) can be performed with relative ease. There are generally no upper or lower ambient or surface temperature limits on the application of thermal spray coatings, although there are practical limits at which workers can properly perform their tasks. In thermal spray, the steel substrate is generally preheated to well above ambient temperatures to drive off any latent moisture and to prevent condensation from forming on the surface. In addition, any ill effects that a cold substrate might have are ameliorated.

e. Maintainability. The future maintainability of the coating system should be considered by the specifier. Some protective coatings are easier to maintain than are others. The specifier should also be cognizant of how maintenance painting is normally achieved, whether by contractor or with in-house labor. In-house labor is usually sufficient for low technology processes that require minimal training and equipment. For example, touch-up painting with brushes or rollers of paints exposed to the atmosphere is readily accomplished with in-house labor. More sophisticated dedicated in-house paint crews can accomplish more complicated work including abrasive blasting and spray application of paints for immersion service. Thermal spray coating and maintenance, because of their specialized nature and relatively high equipment cost, are ordinarily best accomplished by contract. Thermal spray coatings are also more difficult to repair than are most paint coatings. The ease of spot repair of thermal spray metal coatings approximates that of the vinyl paint systems. As with the vinyls, special care must be taken to properly feather the edges of the blast-repaired areas without causing adjacent coating to disbond or lift from the surface. Because of the difficulty affecting appropriate repairs, the thermal spray coating systems, like the vinyls, are generally kept in service until total recoating is needed.

f. Cost. Coating systems are cost effective only to the extent that they will provide the requisite corrosion protection. Cost should be considered only after the identification of coatings that will perform in the exposure environment. Given that a number of coating systems may perform for a given application, the next consideration is the cost of the coating job. Ideally, protective coating systems will always be selected based on life-cycle cost rather than simple installed cost. However, given the realities of budgets, this approach is not always practical.

Therefore, coating systems are sometimes selected on the basis of first or installed cost. Because thermal spray coating systems are almost always more expensive to install than paint systems for a given application, they are often passed over, when, in fact, they can have significantly lower life-cycle costs than paint systems. For additional information on the cost of thermal spray and how to perform cost calculations, refer to paragraph 4-2 of this manual.

5-4. Thermal Spray Selection for Ferrous Metal Surfaces in Fresh Water

All of the thermal spray systems described in CEGS-09971 will perform in freshwater immersion service. The 85-15 zinc-aluminum systems designated as systems 4-Z-A, 5-Z-A, and 6-Z-A are considered to have optimal properties for freshwater service, combining the superior corrosion resistance of zinc and the improved impact and abrasion resistance of aluminum. The more severe the service, the thicker the coating should be, with system 6-Z-A being the recommended choice for highly turbulent ice- and debris-laden waters. System 5-Z-A is the first choice for relatively quiet nonabrasive waters. Seal coats and paint topcoats may be used to add a further degree of protection to the thermal spray coating systems used in freshwater immersion but their use is not considered an absolute necessity. Table 5-1 identifies a number of typical components exposed to freshwater environments and the recommended and preferred thermal spray systems.

5-5. Thermal Spray Selection for Ferrous Metal Surfaces in Seawater

Again, all of the thermal spray systems described in CEGS-09971 will perform in seawater immersion service. Aluminum thermal spray coatings have seen much wider use in marine environments, and they are generally preferred over the zinc-containing coatings for seawater immersion. System 8-A is the recommended thermal spray system for this application. Seal coats and paint topcoats may be used to add a further degree of protection to the thermal spray coating systems used in seawater immersion, but their use is not considered an absolute necessity. Table 5-2 identifies a number of typical components exposed to seawater environments and the recommended and preferred thermal spray systems.

5-6. Thermal Spray Selection for Ferrous Metal Surfaces Exposed to the Atmosphere

Table 5-3 identifies a number of typical components exposed to the atmosphere and the recommended and preferred thermal spray systems.

a. Marine and normal atmospheric exposives. All of the thermal spray systems described in CEGS-09971 will perform in both marine and normal atmospheric exposures. Aluminum thermal spray system 7-A is generally preferred for atmospheric applications where a significant amount of salt can be expected to be deposited on the coated surfaces. These applications include coastal marine structures and bridges exposed to deicing salts. Sealing and topcoating the aluminum thermal spray is optional and may be done for aesthetic reasons or to improve the overall performance of the coating system.

b. Mild atmospheric exposures. Structures in less severe atmospheric environments such as rural areas should be coated with either zinc or zinc-aluminum alloy systems 1-Z and 4-Z-A, respectively. Again, sealing and topcoating the thermal spray coating for such mild atmospheric exposures is not necessary but may be done to increase the service life of the coating system or to alter the appearance.

Table 5-1
Recommended Thermal Spray Systems for Freshwater Immersion

Components	Thermal Spray Systems from CEGS-09971[a]	Sealers
Penstocks, spiral cases, spiral case extensions, draft tube liners, and surge tanks	2-Z, 3-Z, 5-Z-A, **6-Z-A**, 8-A	1 coat C-200A
Crest gates	2-Z, 3-Z, 5-Z-A, **6-Z-A**, 8-A	2 coats V-766e or 1 coat V-766e + 1 coat V-102e or none
Control gates and valves of reservoir outlet works	**6-Z-A**, 8-A	2 coats V-766e or none
Trashracks for water intakes	**6-Z-A**	None
Navigation lock gates and valves	5-Z-A, **6-Z-A**, 8-A	2 coats V-766e or 1 coat V-766e + 1 coat V-102e or none
Navigation dam gates	**6-Z-A**	2 coats V-766e or 1 coat V-766e + 1 coat V-102e or none
Freshwater tanks	2-Z, 3-Z, 5-Z-A, **6-Z-A**, 8-A	1 coat C-200A or none
Equipment for local flood protection projects	Not recommended	
Exterior surfaces of steel hulls	**6-Z-A**, 8-A	2 coats V-766e or 2 coats V-106d or 1 coat V-766e + 2 coats V-103c
Wet interior surfaces of steel hulls	2-Z, **3-Z**, 5-Z-A, 6-Z-A, 8-A	None

[a] System in boldface is preferred.

Table 5-2
Recommended Thermal Spray Systems for Seawater Immersion

Components	Thermal Spray Systems from CEGS-09971[a]	Sealers
Steel piling	**8-A** (from just below mud line to 0.9 m (3 ft) above high-water line)	2 coats V-766e or 1 coat V-766e + 1 coat V-102e
Trashracks for water intakes	6-Z-A, **8-A**	None
Navigation lock gates and valves	6-Z-A, **8-A**	2 coats V-766e or 1 coat V-766e + 1 coat V-102e
Exterior surfaces of steel hulls	6-Z-A, **8-A**	2 coats V-766e or 2 coats V-106d or 1 coat V-766e + 2 coats V-103c

[a] System in boldface is preferred.

c. Severe atmospheric exposures. For severe atmospheric exposures such as industrial areas with acid rain, an aluminum thermal spray system should be used. Thermal spray coatings exposed to corrosive industrial atmospheres or chemical fumes should always be sealed and topcoated.

5-7. Thermal Spray Selection for Ferrous Metal Surfaces Exposed to High Temperatures

Thermal spray coatings of aluminum, zinc, and their alloys will provide excellent long-term performance at temperatures approaching their melting points. The maximum recommended service temperature for zinc systems 2-Z and 3-Z is 60 °C (140 °F). The maximum recommended service temperature for 85-15 zinc-aluminum alloy systems 5-Z-A and 6-Z-A is 315 °C (600 °F). Aluminum thermal spray is an excellent choice for high-temperature applications. System 8-A, described in CEGS-09971, is the preferred system for civil works applications such as stacks of floating plants, where the surface temperature is expected to exceed 400 °C (750 °F). The high-temperature performance of aluminum thermal spray coatings can be further improved by post-heating or fusing of the aluminum to the steel substrate. Fusing is ordinarily accomplished by reheating the aluminum thermal spray coating with oxyacetylene torches. This process fuses the aluminum and steel substrate, creating a metallurgical bond. Coating performance may be further enhanced by applying a seal coat of an aluminum pigmented bitumastic coating. Such coatings can provide corrosion resistance against hot gases at

Table 5-3
Recommended Thermal Spray Systems for Atmospheric Exposures

Components	Thermal Spray Systems from CEGS-09971[a]	Sealers
Exterior surfaces penstocks	2-Z, 3-Z, **5-Z-A**, 6-Z-A, 8-A	1 coat CID A-A-3127
Exterior surfaces of surge tanks	2-Z, **3-Z**, 5-Z-A, 6-Z-A, 8-A	1 coat CID A-A-3127
Service bridges	1-Z, **2-Z**, 3-Z-A, 4-Z-A	2 coats TT-P-38 or 1 coat CID A-A-3127 or 2 coats CID A-A-3132 or 1 coat SSPC Paint 27 + 1 coat A-A-2962
Exterior surfaces of steel tanks	1-Z, **2-Z**, 3-Z-A, 4-Z-A	2 coats TT-P-38 or 1 coat CID A-A-3127 or 2 coats CID A-A-3132 or 1 coat SSPC Paint 27 + 2 coats A-A-2962
Steel decks (nonskid)	**8-A**	1 coat CID A-A-3127 or 2 coats CID A-A-3132
Stacks of floating plants		2 coats TT-P-28
Temp. < 345 °C (650°F)	**6-Z-A**	or 2 coats silicone alkyd
Temp. > 345 to 595 °C (650 to 1100°F)	**8-A**	
Dry interior surfaces of steel hulls	**1-Z**, 2-Z, 3-Z-A, 4-Z-A	None
Steel surfaces in marine atmospheres	3-Z-A, 4-Z-A, 7-A, **8-A**	2 coats TT-P-38 or 1 coat CID A-A-3127 or 2 coats CID A-A-3132 or 1 coat SSPC Paint 27 + 1 coat A-A-2962 or 2 coats V-766e or 1 coat V-7662 + 1 coat V-102e

[a] System in boldface is preferred.

temperatures of up to 870 °C (1600 °F). Thermal spray coatings may not be practical for some high-temperature applications. Thin steel substrates cannot usually be blast-cleaned to create the profile needed for thermal spray coatings without warpage. Steel substrates that cannot be blast- cleaned should be painted rather than thermal spray coated for high-temperature applications. Some typical components exposed to high temperatures and the recommended thermal spray systems can be found in Table 5-3.

5-8. Thermal Spray Selection for Zebra Mussel Protection

The zebra mussel is a freshwater bivalve that colonizes hard substrates. When present in sufficient densities, the zebra mussel can impact the performance of civil works structures and floating plants. Zebra mussels can impair structure performance by occluding narrow openings and small conduits such as may be found in condenser tubes, trashracks, sea chests, fire suppression systems, etc. Zebra mussels can also reduce the efficiency of floating plants and power plants by causing hydraulic drag. Coatings are one means of preventing the attachment of zebra mussels to Corps structures. Thermal spray coatings containing zinc and/or copper have been found to be effective deterrents. Juvenile mussels will not settle on these metallic surfaces because small amounts of zinc and copper leaching into the water from the coating act as deterrents. The mussels are not actually killed but rather they sense the toxic chemicals in the water and do not attach themselves to these surfaces. Thermal spray coatings containing copper such as brass, aluminum bronze, and pure copper should not be applied to steel substrates because they will not protect the steel from corrosion and may make it worse. Copper and brass thermal spray coatings can however be applied to concrete substrates to prevent zebra mussel fouling. Zinc and 85-15 zinc-aluminum alloy coatings are a better choice for controlling zebra mussels on steel as they also serve as anticorrosive coatings. Because of their lower material costs zinc and 85-15 zinc-aluminum are probably better choices for concrete as well, even though they are slightly less effective than brass and copper. Zinc-containing coatings do not need to be registered with the EPA. Zinc is a relatively weak aquatic toxin and is an even weaker mammalian toxin, and as such, no effects on nontarget organisms should be anticipated with its prophylactic use as a zebra mussel deterrent.

5-9. Thermal Spray Coatings for Cathodic Protection of Reinforcing Steel in Concrete

Steel reinforced concrete structures such as bridges, parking decks, and piers are prone to chloride-induced corrosion. Chloride ions present in deicing salts and marine atmospheres penetrate the concrete monolith with time. The normally passivated steel rebar will begin to corrode when enough salt accumulates at the steel-concrete interface. The steel corrosion products, being more voluminous than the steel itself, cause the concrete to crack and spall. Failures of reinforced concrete systems can be very expensive to repair and are difficult to prevent. Various approaches to preventing this phenomenon have been tried with varying degrees of success including epoxy coated rebar, galvanized rebar, special concrete admixtures, and sealing the concrete to prevent chloride penetration. A more effective method of preventing chloride-induced failures is the use of cathodic protection. Zinc thermal spray can been used as either a consumable galvanic anode or as a conductive anode in an impressed current system. In the impressed current system, rectifiers are used to supply current to the conductive zinc anode via electrical connectors which are attached to the concrete. The zinc thermal spray coating is itself applied directly to the concrete. Anode design and current density are important to the overall effectiveness of the cathodic protection system. At this time there is no guidance within the Corps on the use of zinc metallized cathodic protection systems. For additional information on cathodic protection refer to EM 1110-2-2704, "Cathodic Protection Systems for Civil Works Projects," TM 5-811-7, "Electrical Design, Cathodic Protection," and ETL 1110-3-474, "Cathodic Protection."

5-10. Thermal Spray Nonskid Coatings

Nonskid coatings are sometimes used to prevent or reduce slip hazards. Aluminum thermal spray coatings have been used successfully on metal substrates to prevent corrosion and impart nonskid properties. Historically, paint coatings have been used for nonskid applications. However, because of their greater hardness and roughness, aluminum thermal spray coatings are superior for many nonskid applications. The nonskid coating system is achieved by first applying aluminum thermal spray system 8-A. An additional spray pass of aluminum is then applied using reduced atomization air pressure. The lower air pressure allows for the deposition of larger spray particles that produce a rougher surface. Nonskid coatings should be sealed with thin film epoxies. Aliphatic polyurethanes can be used for durable striping if desired. Some common applications for nonskid coatings and the recommended thermal spray coatings systems may be found in Table 5-3.

5-11. Thermal Spray Coatings for Cavitation/Erosion Protection

a. Cavitation repair and mitigation. For hydraulic components subject to low cavitation environments, Stellite 6 applied by the HVOF spray process may be used to mitigate and repair cavitation damage. The coating is applied to a thickness of 500 μm (0.020 in.) over blast-cleaned weld metal overlay for cavitation repair or directly to the blast-cleaned steel component for mitigating cavitation damage. The coating serves as a sacrificial element with improved wear resistance. With proper maintenance intervals, the coating may be replaced periodically at approximately a third of the cost of additional maintenance by weld overlay. Turbine draft tube liners and pump impellers are good candidates for the use of Stellite 6 coatings for cavitation repair and mitigation.

b. Dissimilar metals corrosion. Weld overlay repair of hydraulic components is generally performed using a stainless steel material over a carbon steel substrate. The two metals have different electrochemical potentials, and, therefore, galvanic corrosion will occur to the mild steel adjacent to the boundary of the two metals. The corrosion may be exacerbated by the erosion taking place in the cavitating environment. Stellite 6 applied by the

HVOF spray process may be used to mitigate corrosion of hydraulic components subject to dissimilar metals corrosion in erosive environments. The coating is applied to a thickness of 500 μm (0.020 in.) over the entire component, including the blast-cleaned weld metal overlay. Stellite 6 acts as an expendable wear resistance coating. Turbine draft tube liners and pump impellers are good candidates for the use of Stellite 6 coatings for corrosion mitigation. Arc plasma sprayed alumina titania ceramic powder coatings can also be used to improve and restore dimensions of pump shafts and bearings and to provide an erosion-corrosion resistant coating for impellers and interior surfaces of the casing. Alumina titania is a ceramic coating and is not subject to galvanic corrosion.

c. For additional information on methods and materials for thermal spray cavitation/erosion protection see USACERL Technical Report 97/118, "Cavitation- and Erosion-Resistant Thermal Spray Coatings," and MIL-STD-1687A(SH).

5-12. Thermal Spray Coatings for Partially Submerged Structures

Certain components of hydraulic structures may be only partially submerged. Structural components that are partially immersed in either seawater or fresh water should be coated with the appropriate thermal spray system for immersion. The aerial exposed portions of the structural component should be coated with the same thermal spray material of the same or lesser thickness. Alternatively, the aerial exposures may be protected with just a paint system. If this method is selected, the thermal spray sealer and the paint topcoat should be the same material. For example, a miter gate partially immersed in seawater could be metallized with aluminum system 8-A and sealed with two coats of vinyl paint V-766e. Above the waterline, the gate could be coated with aluminum thermal spray system 7-A and sealed with two coats of V-766e. Alternatively, paint system 5-E-Z, described in CEGS-09965 and consisting of a vinyl zinc-rich primer (VZ-108d) and multiple coats of gray and white vinyl (V-766e), could be used to protect the atmospherically exposed portions of the gate. Both the paint system and the aluminum thermal spray system would provide excellent protection at a reduced cost. Dissimilar thermal spray metals should never be applied to the same structural component because one of the materials may corrode preferentially to protect the other metal.

Chapter 6
Surface Preparation

6-1. Introduction

Thermal spray coatings require a very clean surface that is free of oil, grease, dirt, and soluble salts. Surface contaminants must be cleaned with solvents prior to removal of mill scale, corrosion products, and old paint by abrasive blasting.

a. Surface preparation is the single most important factor in determining the success of the corrosion protective thermal spray coating system. Abrasive blasting or abrasive blasting combined with other surface preparation techniques is used to create the necessary degree of surface cleanliness and roughness.

b. The principal objective of surface preparation is to achieve proper adhesion of the thermal spray coating to the steel substrate. Adhesion is the key to the success of the thermal spray coating.

c. The purpose of surface preparation is to roughen the surface, creating increased surface area for mechanical bonding of the thermal spray coating to the steel substrate. The roughening is typically referred to as the anchor pattern or profile. The profile is a pattern of peaks and valleys that is etched onto the steel when high-velocity abrasive blast particles impinge upon the surface.

d. Surface cleanliness is essential for proper adhesion of the thermal spray coating to the substrate. Thermal spray coatings applied over rust, dirt, grease, or oil will have poor adhesion. Premature failure of the thermal spray coating may result from application to contaminated substrates.

6-2. Solvent Cleaning (SSPC-SP 1)

Solvent cleaning (SSPC-SP 1) is a procedure for removing surface contaminants, including oil, grease, dirt, drawing and cutting compounds, and soluble salts, from steel surfaces by means of solvents, water, detergents, emulsifying agents, and steam. These methods are not designed to remove mill scale, rust, or old coatings. Ineffective use of the solvent cleaning technique may spread or incompletely remove surface contaminants. Three common methods of solvent cleaning are water washing, steam cleaning, and cleaning with hydrocarbon solvents.

a. Water cleaning. Low-pressure water cleaning, up to 34,000 kPa (5000 psi)), is an effective means of removing dirt and soluble salt contamination. When used with a detergent or emulsifying agent, the method can be used to remove organic contaminants such as grease and oil. Thorough rinsing with clean water will ensure complete removal of the cleaning agent. If an alkaline cleaner is used, the pH of the cleaned surface should be checked after the final rinse to ensure that the cleaning agent has been completely removed.

b. Steam cleaning. Steam cleaning is an effective means of removing dirt, salt, oil, and grease from both coated and uncoated substrates. The method employs a combination of detergent action and high-pressure heated water (138 °C (280 °F) to 149 °C (300 °F) at 11.3 to 18.9 ℓ/min (3 to 5 gpm)). Thorough rinsing with steam or water should be used to remove any deposited detergent.

c. Hydrocarbon solvent cleaning. Hydrocarbon solvents used to remove grease and oil are typically petroleum-based distillates as described by ASTM D235 "Standard Specification for Mineral Spirits (Petroleum Spirits) (Hydrocarbon Dry Cleaning Solvent)." Type I - regular (Stoddard Solvent) with a minimum flash point of 38 °C (100 °F). Type II - High Flash Point mineral spirits with a minimum flash point of 60 °C (140 °F) should be used when ambient temperatures exceed 35 °C (95 °F). Aromatic solvents such as xylene and high flash aromatic naphtha 100 or 150 (ASTM D 3734 Types I and II) are sometimes used when a stronger solvent is needed. The use of aromatic hydrocarbons should be limited because of their generally greater toxicity. Solvent cleaning with hydrocarbon solvents is typically accomplished by wiping the surface with solvent soaked rags. Rags should be changed frequently to afford better removal and to prevent spreading and depositing a thin layer of grease or oil on the surface

6-3. Abrasive Blast Cleaning

Abrasive blasting is performed in preparation for thermal spray after the removal of surface contaminants by solvent cleaning. Abrasive blasting is conducted to remove mill scale, rust, and old coatings, as well as to provide the surface profile necessary for good adhesion of the thermal spray coating to the substrate. Conventional abrasive blast cleaning is accomplished through the high-velocity (724 km/h (450 mph)) propulsion of a blast media in a stream of compressed air (620 to 698 kPa (90 to 100 psi)) against the substrate. The particles' mass and high velocity combine to produce kinetic energy sufficient to remove rust, mill scale, and old coatings from the substrate while simultaneously producing a roughened surface. The Society for Protective Coatings (SSPC) and the National Association of Corrosion Engineers (NACE) have published standards for surface cleanliness. These standards and an SSPC supplemental pictorial guide provide guidelines for various degrees of surface cleanliness. Only the highest degree of cleanliness, SSPC-SP-5 "White Metal Blast Cleaning" or NACE #1, is considered acceptable for thermal spray coatings. Paragraph 6-7 discusses these standards in greater detail. Abrasive blast cleaning may be broadly categorized into centrifugal blast cleaning and air abrasive blast cleaning. Air abrasive blast cleaning may be further subdivided to include open nozzle, water blast with abrasive injection, open nozzle with a water collar, automated blast cleaning, and vacuum blast cleaning. Open nozzle blasting is the method most applicable to preparation for thermal spray coating.

a. Equipment. An open nozzle abrasive blast-cleaning apparatus consists of an air compressor, air hose, moisture and oil separators/air coolers and dryers, blast pot, blast hose, nozzle, and safety equipment.

(1) Air compressor. The air compressor supplies air to the system to carry the abrasive. Production rate depends on the volume of air that the compressor can deliver. A larger compressor can supply more air and can therefore sustain operation of more blast nozzles or larger blast nozzle diameters.

(2) Air hose. The air hose supplies air from the compressor to the blast pot. The air hose should be as short, with as few couplings, and as large of diameter as possible to optimize efficiency. The minimum inside diameter (i.d.) should be 31.75 mm (1.25 in.) with measurements of 50.8 to 101.6 mm (2 to 4 in.) i.d. being common.

(3) Moisture and oil separators/air coolers and dryers. If not removed, moisture from the air and oil mists from the compressor lubricants may contaminate the abrasive in the blast pot and subsequently the surface being cleaned. Oil/moisture separators are used to alleviate this problem. The devices should be placed at the end of the air hose as close to the blast pot as possible. Separators are typically of the cyclone type with expansion air chambers and micron air filters. Air coolers/dryers are commonly used to treat the air produced by the compressor.

(4) Blast pot. Most blast pots used for large blasting projects are of the gravity-flow type. These machines maintain equal pressure on top and beneath the abrasive. The typical blast pot consists of air inlet and outlet valves, a filling head, a metering valve for regulating abrasive flow, and a hand hole for removing foreign objects from the pot chamber. For large jobs, the pot should hold enough media to blast for 30 to 40 min. For continuous production, a two-pot unit can be used, allowing one pot to be filled while the other operates.

(5) Blast hose. The blast hose carries the air-media mixture from the blast pot to the nozzle. A rugged multi-ply hose with a minimum 31.75-mm (1.25-in.) i.d. is common. A lighter, more flexible length of hose called a whip is sometimes used for added mobility at the nozzle end of the blast hose. Maximum blast efficiency is attained with the shortest, straightest blast hoses. Blast hoses should be coupled with external quick-connect couplings.

(6) Blast nozzle. Blast nozzles are characterized by their diameter, material, length, and shape. Nozzle sizes are designated by the inside diameter of the orifice and are measured in sixteenths of an inch. A 3/16-in.-diam orifice is designated as a No. 3 nozzle. The nozzle diameter must be properly sized to match the volume of air available. Too large an orifice will cause pressure to drop and production to decrease. Too small an orifice will not fully utilize the available air volume. The nozzle size should be as large as possible while still maintaining an air pressure of 620 to 689 kPa (90 to 100 psi) at the nozzle. Blast nozzles may be lined with a variety of different materials distinguished by their relative hardness and resistance to wear. Ceramic and cast iron lined nozzles have the shortest life. Tungsten and boron carbide are long lived nozzles. Nozzles may be either straight bore or venturi-type. The venturi nozzle is tapered in the middle, resulting in much higher particle velocities. Venturi nozzles have production rates 30 to 50 percent higher than straight bore nozzles. Long nozzles, 127 to 203 mm (5 to 8 in.), will more readily remove tightly adherent rust and mill scale and increase production rates. Worn nozzles can greatly decrease production and should be replaced as soon as they increase one size (1/16 in.).

b. Blast cleaning techniques. Proper blasting technique is important in order to accomplish the work efficiently with a high degree of quality. The blast operator must maintain the optimal standoff distance, nozzle angle, and abrasive flow rate. The best combination of these parameters is determined by an experienced blaster on a job-to-job basis.

(1) The blaster should balance the abrasive and air flows to produce a "bluish" colored abrasive airstream at the nozzle which signals the optimum mix. Blasters often use too much abrasive in the mix which results in reduced efficiency. The mix is adjusted using the valve at the base of the blast pot.

(2) The nozzle-to-surface angle should be varied to achieve the optimal blast performance for the given conditions. Rust and mill scale are best removed by maintaining a nozzle-to-surface angle of 80 to 90 deg. A slight downward angle will direct dust away from the operator and improve visibility. The best nozzle-to-surface angles for removing old paint are 45 to 70 deg. The final blast profile should always be achieved with a nozzle-to-surface angle of 80 to 90 deg.

(3) Standoff, or nozzle-to-surface distance, will also affect the quality and speed of blast cleaning. The lower the standoff distance, the smaller the blast pattern will be, and the longer it will take to cover a given area. However, close standoff distances allow for more energy to be imparted to the surface, allowing for the removal of more tenacious deposits such as mill scale. A standoff distance of as little as 153 mm (6 in.) may be necessary for the removal of tight mill scale and heavy rust deposits. Higher standoff distances, on the order of 457 mm (18 in.), are more efficient for the removal of old loosely adherent coatings.

c. Abrasive media type and selection. The selection of the proper blast media type and size is critical to the performance of the thermal spray coating. Blast media that produce very dense and angular blast profiles of the appropriate depth must be used. An angular blast media must always be used. Rounded media such as steel shot, or mixtures of round and angular media will not produce the appropriate degree of angularity in the blast profile. The adhesion of thermal spray coatings can vary by an order of magnitude as a function of profile shape and depth. Thermal spray coatings adhere poorly to substrates prepared with rounded media and may fail in-service by spontaneous delamination. Hard, dense, angular blast media such as aluminum oxide, iron oxide, and angular steel grit are needed to achieve the depth and shape of blast profile necessary for good thermal spray adhesion. Steel grit should be manufactured from crushed steel shot conforming to SAE J827. Steel grit media composed of irregular shaped particles or mixtures of irregular and angular particles should never be used. Only angular steel grit should be used. New steel grit should conform to the requirements of SSPC-AB 3, "Newly Manufactured or Re-Manufactured Steel Abrasives." Various hardnesses of steel grit are available but generally grit with Rockwell C hardness in the range of 50 to 60 is used. Harder steel grit (Rockwell C 60 to 66) may also be used provided that the proper surface profile is obtained. Steel shot and slag abrasives composed of all rounded or mixed angular, irregular, and rounded particles should never be used to profile steel for thermal spraying. Table 6-1 shows the recommended blast media types as a function of the thermal spray process and coating material.

Table 6-1
Recommended Blast Media for Thermal Spray Surface Preparation

Thermal Spray Material	Thermal Spray Process	Blasting Media
Aluminum, zinc, and 85-15 zinc-aluminum	Wire flame spray	Aluminum oxide Angular steel grit
Aluminum, zinc, and 85-15 zinc-aluminum	Arc spray	Aluminum oxide Angular steel grit, angular iron oxide
Aluminum and zinc	Powder flame spray	Aluminum oxide Angular steel grit

d. Blast profile. Thermal spray coatings are generally more highly stressed than paint coatings and as such require a deeper blast profile to dissipate the tensile forces within the coating. In general, the greater the thickness of thermal spray coating being applied, the deeper the blast profile that is required. The minimum recommended blast profile for the thinnest coatings of zinc and 85-15 zinc-aluminum (100 to 150 μm (0.004 to 0.006 in.)) is 50 μm (0.002 in.). Thicker coatings of zinc and 85-15 zinc-aluminum, 250 μm (0.010 in.) or greater, require a minimum 75-μm (0.003-in.) profile. A 125-μm-(0.005-in.-) thick aluminum coating requires a minimum surface profile of 50 μm (0.002 in.), and a 250-μm (0.010-in.) aluminum coating requires a minimum 62.5-μm (0.0025-in.) profile. The specifier should specify the maximum and minimum surface profile required for the thermal spray coating. The maximum profile for thicker thermal spray coatings should not exceed approximately a third of the total average coating thickness. As a general rule, the maximum blast profile should be 25 μm (0.001 in.) greater than the specified minimum profile depth. Table 6-2 shows the recommended minimum and maximum blast profile depths for the thermal spray systems described in CEGS-09971. Table 6-3 shows typical surface profiles produced by selected steel grit abrasives.

e. Blasting with reusable media.

(1) Durable reusable blast media are now commonplace in industrial maintenance coating. The extensive use of blasting enclosures used to contain paint and blasting debris has made the use of reusable abrasives more economical for field applications. Reusable abrasives require the use of a reclaiming and recycling system. Commonly used recyclable abrasives that are appropriate for use in preparing steel for thermal spray include steel grit, aluminum oxide, and iron oxide. Iron oxide, aluminum oxide, and steel grit may be reused about 4, 6 to 8, and 100+ times, respectively. Because it is more economical, steel grit is much more likely to be used for fieldwork than are other reusable media.

(2) Control of the working mix of abrasive is critical to maintaining the quality of the blast. The working mix should be sampled frequently and subjected to a sieve analysis to determine the particle size distribution. The distribution can be used to determine the frequency of make-up additions to the working mix. If the mix is allowed to become depleted of larger particle sizes, then the required surface profile depth will not be achieved.

Table 6-2 Recommended Surface Profiles for Thermal Spray Systems			
Thermal Spray System Designation	Thermal Spray Material	Minimum/Average Thermal Spray Thickness, microns (in.)	Minimum/Maximum Surface Profile, microns (in.)[a]
1-Z	Zinc	125/150 (0.005/0.006)	50/75 (0.002/0.003)
2-Z	Zinc	250/300 (0.010/0.012)	62.5/87.5 (0.0025/0.0035)
3-Z	Zinc	350/400 (0.014/0.016)	75/100 (0.003/0.004)
4-Z-A	85-15 zinc-aluminum	125/150 (0.005/0.006)	50/75 (0.002/0.003)
5-Z-A	85-15 zinc-aluminum	250/300 (0.010/0.012)	62.5/87.5 (0.0025/0.0035)
6-Z-A	85-15 zinc-aluminum	350/400 (0.014/0.016)	75/100 (0.003/0.004)
7-A	Aluminum	100/125 (0.004/0.005)	50/75 (0.002/0.003)
8-A	Aluminum	200/250 (0.008/0.010)	62.5/87.5 (0.0025/0.0035)

[a] As measured by ASTM D4417, Method C (replica tape). This method measures the average maximum peak to valley height.

Table 6-3 Typical Surface Profiles for Selected Steel Grit Abrasives[a]	
Steel Grit Size	Surface Profile, 0.001 in.[b]
G50	1.6 ± 0.3
G40	2.4 ± 0.5
G25	3.1 ± 0.7
G14	5.1 ± 0.9

[a] Steel grit is crushed steel shot conforming to SAE J827. Grit hardness is 55-60 Rockwell C.
[b] As measured by ASTM D4417, Method C (replica tape). This method measures the average maximum peak to valley height.

This may lead to poor thermal spray coating adhesion and premature failure. Different abrasive materials wear out at different rates, usually given as consumption rates in pounds per hour. In theory, new abrasive can be added at a rate equal to the consumption rate. A continuous system of abrasive replenishment is the preferred method of maintaining the proper working mix and should be required on all jobs where recyclable abrasives are used. Maintaining a uniform working mixture also requires the removal of all particles below a given minimum size, which is the smallest size that is still effective in the cleaning operation. Recycled steel grit should meet the cleanliness requirements of SSPC-AB 2.

(3) The size and type of abrasive blast system used with recyclable media must be properly selected. Contractors employing undersized equipment or an inappropriate type of equipment for the job are not likely to produce a quality surface with the correct profile. In general, the contractor should employ midsize, 9 to 13.5 metric tons per hour (10 to 15 tons per hour), or large, >13.5 metric tons per hour (15 tons per hour), blast media recycling systems. Midsize and large blast media recycling systems typically have better multistage abrasive cleaning systems that remove dust, debris, and small particles.

f. Centrifugal blast cleaning. Centrifugal blast cleaning is commonly used in fabrication shops. The method is generally faster and more economical than open abrasive blasting. The method involves the conveying of the steel through a blast cabinet or enclosure where high-speed rotating wheels with blades propel abrasive particles at the steel. The blasting debris falls to the bottom of the chamber where it is reclaimed, cleaned, and then recycled. The degree of cleanliness achieved is determined by the abrasive velocity and the conveyor speed. Steel shot is usually used in centrifugal blast machines. For thermal spray coatings, a subsequent profiling blast using an angular media is required to achieve the desired blast profile depth and angularity. Centrifugal blast cleaning machines are now available for fieldwork as well, but their use is not widespread.

g. Cleaning after blasting. Cleanliness after abrasive blasting is important. Any remaining traces of spent abrasive or other debris must be blown, swept, or vacuumed from the surface prior to thermal spray application.

A hard-to-see layer of abrasive dust may adhere to the substrate by static electric charge and must be removed. The thermal spray applicator may accomplish this by triggering just the compressed air from the flame or arc gun. Scaffolding, staging, or support steel above the thermal spray coating area must also be cleaned prior to application to prevent debris from falling onto the surfaces to be coated. Blasting and thermal spray should not occur simultaneously unless the two operations can be adequately isolated to prevent contamination of the thermal spray surfaces.

h. Time between blasting and thermal spraying. After completion and inspection of the final profiling blast, the steel substrate should be thermal sprayed as soon as possible. A maximum period of 4 hr is generally allowed to elapse between the completion of blast cleaning and thermal spraying. This period should allow adequate time for the changeover from blasting to thermal spraying. Thermal spray should commence prior to the appearance of any visible rust bloom on the surface. Foreign matter such as paint overspray, dust and debris, and precipitation should not be allowed to contact the prepared surfaces prior to thermal spraying. Under no circumstances should the application of thermal spray be allowed on rerusted or contaminated surfaces. In some cases it may be possible to apply only a single spray pass or some other fraction of the total thermal spray system within 4 hr of blasting. This single layer must cover the peaks of the surface profile. The partial coating is intended to temporarily preserve the surface preparation. Before applying additional sprayed metal to the specified thickness, the first layer of coating should be visually inspected to verify that the coating surface has not been contaminated. Any contamination between coats should be removed before any additional material is applied. The remaining coating should be sprayed to achieve the specified thickness as soon as possible. In some cases it may be possible to hold the surface preparation for extended periods using specially designed dehumidification (DH) systems. These systems supply dry air to a blast enclosure or other contained air space. The dry air prevents the reappearance of rust for extended periods of time and allows for thermal spray jobs to be staged in a different fashion. DH systems may be particularly useful for jobs in very humid environments, which are typical of many Corps facilities during the spring through fall maintenance season. These areas typically have dense morning fog and hot humid afternoons. Holding the quality of blast needed for thermal spray coatings would be difficult under such conditions without the use of DH.

6-4. Minimizing Surface Preparation Costs

a. Strip blasting. Combined methods of surface preparation may be used to reduce the overall cost of surface preparation. The most common method of combined surface preparation is the use of strip blasting with an inexpensive abrasive. Strip blasting may be used to remove aged coatings or to remove tightly adherent mill scale. Steel shot is particularly effective for removing mill scale. Steel, copper, nickel, and coal slags and garnet and zircon abrasives may all be used to remove old coatings. Silica-containing materials, including silica and mineral sands, glass beads, flint, and novaculite, should be avoided because of health issues surrounding the use of abrasives containing silica.

b. Profile blasting. Profile blasting must be performed after strip blasting in all cases. It is not necessary for profile blasting to occur before light rerusting of the surface occurs. The strip blasted surface should, however, be protected from the deposition of contaminants such as grease and oil that may not be removed during the final profiling blast. Under some circumstances, it may be economical to perform strip blasting using ultrahigh-pressure water jetting. This method uses high pressure water, >170,000 kPa (25,000 psi), to remove old coatings. The method will not, however, remove mill scale or tightly adherent rust. The combination of ultrahigh-pressure water jetting and profile blasting probably will not be economical for surfaces with mill scale. Pressurized water cleaning systems are used frequently in shipyards and are most likely to be encountered in strip and profile applications at these facilities.

6-5. Preparing Heat-Affected Zones

Heat-affected zones associated with steel welding and cutting operations can produce surfaces that are difficult to clean and thermal spray. Welding and cutting operations produce enough heat to anneal, or harden, the surface of the steel. In some cases, the steel may be so hard as to prevent adequate profiling during abrasive blasting. It is recommended that all heat-affected zones first be ground with a disk wheel grinder prior to profile blasting. For example, the hardened or carburized layer on a flame-cut girder flange should be ground off before abrasive blasting for profile. Weld spatter not removed by blasting should be removed with impact or grinding tools, and the areas should be reblasted prior to thermal spraying.

6-6. Preparing Pitted Steel and Edge Surfaces

a. Pitted Steel. Heavily corroded, deeply pitted surfaces are difficult to prepare for thermal spray coating. Wide, shallow pits do not pose any particular problem, but deep and irregular shaped pits can pose a problem. Pits with an aspect ratio of greater than unity (as deep as they are wide) should be ground with an abrasive disk or other tool prior to blasting. Pits with sharp edges, undercut pits, and pits with an irregular horizontal or vertical orientation must be ground smooth prior to abrasive blasting. Grinding does not need to level or blend the pit with the surrounding steel but should smooth all the rough and irregular surfaces to the extent necessary to allow the entire surface of the pit to be blasted and coated. Nozzle-to-surface angles of 80 to 90 deg are optimal for cleaning pits. Heavily pitted steel on bridges or in other environments where soluble salt contamination is likely should be cleaned with high-pressure water after grinding to ensure that salt contaminants are removed from the pits.

b. Edge Surfaces. Sharp edges also present problems in achieving adequate surface preparation and coating. As a general rule, all sharp edges should be ground prior to blasting to a uniform minimum diameter of 3 mm (1/8 in.).

6-7. Surface Preparation Standards and Specifications

SSPC and NACE have developed blast cleaning standards and specifications for steel surfaces. More detailed information is available on blast cleaning specifications in the SSPC "Painting Manual, Volume 2, Systems and

a. SSPC-SP 5 or NACE #1. SSPC-SP 5 and NACE #1 describe the condition of the blast-cleaned surface when viewed without magnification as free of all visible oil, grease, dust, dirt, mill scale, rust, coating, oxides, corrosion products, and other foreign matter.

b. SSPC VIS 1-89. SSPC VIS 1-89 supplements the written blast standards with a series of photographs depicting the appearance of four grades of blast cleaning over four initial grades of mill scale and rust. The last two pages of the standard depict a white metal blast-cleaned substrate achieved with three different types of metallic abrasives and three types of nonmetallic abrasives. The resulting surfaces have slight color and hue differences caused by the type of media used.

Chapter 7
Thermal Spray Coating Application

7-1. Introduction

a. Thermal spray coatings for corrosion protection of steel may be applied by several different methods, including wire flame spray, arc spray, and powder flame spray. Each process has its inherent advantages.

b. The best application process for a given job depends on the coating material to be applied and the size of the job. Contracts should permit the contractor to select the application process or processes to be used.

c. The ambient conditions under which the thermal spray coating will be applied must be within the specified range.

d. The applicator must employ the proper application techniques, and the equipment must be properly set up and operated within the manufacturer's parameters to ensure the application of a quality coating.

e. The sequencing of the application must occur in a timely manner to ensure that the receiving surfaces are still clean and free of rust bloom.

7-2. Ambient Conditions Required for Thermal Spray

a. Air temperature, humidity, and dew point. Generally speaking, there are no ambient air temperature limitations on the application of thermal spray coatings. Unlike paint coatings, thermal spray is not affected by extremes of temperature. Although there are no theoretical limits within the normal ambient temperature range on the application of thermal spray coatings, in practice there are limits within which the applicator will be effective and safe. Because thermal spray coatings are typically sealed with paint-type coatings, there may be practical limits based on the temperature requirements of the sealer. Most epoxy-type sealers can be applied when ambient temperatures are between 7 and 32 °C (45 and 90 °F). With special precautions, vinyl-type sealers can be applied at temperatures between -18 and 38 °C (0 and 100 °F). However, the application of sealers and paints can often be delayed, provided the thermal spray coated surface remains clean and dry. As with painting, the relative humidity and dew point can affect the quality of the applied coating and may also affect improperly stored or packaged thermal spray feedstock materials. In very humid environments, blast cleaned steel may rerust or exhibit a rust bloom more rapidly than under normal ambient conditions. Thermal spray coatings should never be applied after the appearance of rust bloom on the surface. The dew point is the temperature at which moisture will condense. Condensation and the formation of rust bloom on the steel surface are very likely to occur at or near the dew point. Because of this problem, it is required that the ambient temperature be at least 3 °C (5 °F) above the dew point. Temperature, humidity, and dew point cause problems if thermal spray feedstock is not properly stored. All of the active metal wires oxidize. The oxide film can cause feed problems in both flame and arc equipment. Extreme temperature changes may also cause zinc and zinc-aluminum alloy wire to recrystallize and become brittle. Powder storage is even more critical than wire storage. Moisture in the powder will have an adverse impact on flow in the powder feed systems. Thermal spray wires and powders should be securely sealed and protected from moisture intrusion to prevent oxidation of the material.

b. Steel temperature. Temperature changes in heavy structural steel often lag behind changes in the ambient temperature. This may cause particular problems in the morning hours when the temperature is near the dew point. Steel that is in contact with soil or water may also remain colder than the surrounding air temperature, causing problems with condensation and rerusting of the steel. Provided that the surface has not rerusted, the temperature of the steel is of little consequence. When using combustion spray processes which employ a hydrocarbon fuel gas the surface should be preheated to above 121 °C (250 °F) to prevent condensation of water from the flame on the surface as the coating is being applied. The water in the flame is a by-product of the combustion reaction. Preheating the surface is ordinarily accomplished using the thermal spray gun with the material feed turned off, as a torch. Approximately 0.1 to 0.2 m^2 (1 to 2 ft^2) of surface area should be preheated at a time. Arc spraying does not require preheating of the substrate.

7-3. Thermal Spray Application Techniques

Proper spray technique is critical to the success of thermal spray coatings. Poor spray technique may result in early coating failure due to poor coating adhesion or cohesion, excessive coating porosity, or a high oxide content. Poor spray technique may also result in highly variable coating thicknesses, including areas that are deficient.

a. Spray pattern. Manually applied thermal spray coatings should be applied in a block pattern measuring approximately 60 cm (24 in.) on a side. Each spray pass should be applied parallel to and overlapping the previous pass by approximately 40 percent. Successive spray coats should be applied at right angles to the previous coat until the desired coating thickness is achieved. Approximately 50 to 75 μm (0.002 to 0.003 in.) of coating should be applied per spray pass. In no case should less than two spray coats applied at right angles be used to achieve the specified coating thickness. This procedure is designed to produce the most uniform coating thickness of the best possible quality. A larger block pattern may cause the applicator to overreach, resulting in coating nonuniformity. Wire flame spray guns produce a small round spray pattern, allowing the applicator to maintain the same wrist orientation (vertical) on each spray coat. Alternatively, the wire flame spray applicator may apply the second spray coat with the wrist oriented in the horizontal plane. Arc and powder flame spray guns with fan air caps generally produce a much larger oval shaped spray pattern. The powder flame and arc spray operators generally must spray the second coat with the wrist oriented 90 deg to that used for the first spray coat. The arc and powder flame spray guns should never be used to apply coating moving in a direction parallel to the long access of the oval shaped spray pattern. Most powder flame and arc spray guns may be fitted with optional air caps that will produce different size and shape spray patterns. These fittings allow the operator to apply a uniform coating to components with complex shapes. The large oval fan pattern is best suited to spraying large flat substrates. Small round spray patterns are better suited for coating complex shapes or small objects. Table 7-1 shows the nominal spray widths produced by several types of spray guns.

b. Standoff distance. Standoff distance is dependent on the type and source of spray application equipment used. The maximum standoff distance for most equipment is generally on the order of 15 to 25 cm (6 to 10 in.). Table 7-1 shows nominal standoff distances for various types of equipment. Excessive standoff distance will produce a more porous and oxidized coating with reduced cohesion and adhesion. The higher porosity may be attributed to the greater degree of cooling and the lower velocity that the thermal spray particles experience prior to impact. Adhesion is directly proportional to the kinetic energy of the spray particles, and the kinetic energy varies as the square of the particle velocity. The cooler, slower impacting particles will not adhere as well to each other or to the substrate, resulting in a weaker, less adherent coating. Excessive standoff distance may occur because the applicator is not adequately familiar with the requirements of the equipment or due to fatigue or carelessness. Increased standoff distance may result from the applicator's arm or wrist arcing during application. It is very important that the applicator's arm moves parallel to the substrate to maintain a consistent standoff distance. Holding the thermal spray gun too close to the surface may result in poor coverage and erratic coating thicknesses because of the reduced size of the spray pattern. It is very important that the applicator's arm moves parallel to the substrate while maintaining a constant standoff distance.

Table 7-1
Nominal Standoff Distances and Spray Widths

Thermal Spray Process	Standoff Distance cm (in.)	Spray Width with Regular Air Cap, cm (in.)	Spray Width with Fan Air Cap cm (in.)
Wire flame spray	12.5-17.5 (5-7)	2 (3/4)	Not available
Powder flame spray	20-25 (8-10)	5 (2)	7.5-10 (3-4)
Wire arc spray	15-20 (6-8)	4 (1 1/2)	7.5-10 (3-4)[a]

[a] Newer high production rate wire arc systems may have fan-type air caps that produce spray deposit patterns as wide as 15-25 cm (6-10 in.).

c. Spray angle. The gun-to-surface angle is important because of the generally greater distances that the spray particles travel prior to striking the substrate, producing a situation analogous to the excessive standoff distances already discussed. Porosity, oxide content, and adhesion are strongly affected by spray angle. In some cases, it may be necessary for the applicator to spray at less than 90 deg because of limited access to the surface. In no case should the applicator spray at an angle less than 45 deg. Spray extensions are available from some equipment manufacturers that allow better access to difficult-to-spray areas. A good spray technique consists of the applicator maintaining the spray gun perpendicular (90 deg) or near perpendicular (0 ± 5 deg) at all times.

7-4. Thermal Spray Equipment Operation

Each type and source of thermal spray equipment should be set up and operated in accordance with the manufacturer's recommended procedures. Spray parameters should be optimized primarily for coating quality and secondarily for production rate.

a. Wire and powder flame spray.

(1) Oxygen and fuel gas flow rates. The use of oxygen and fuel gas flow meters allows for the best control of the flame and thus higher spray rates. Under conditions of continuous use, the actual oxygen and fuel gas flow rates and pressures should remain nearly constant and, ordinarily, should not deviate from the set values by more than 5 percent.

(2) Atomization air pressure. Compressed air should be oil- and water-free. Accurate air regulation is necessary to achieve uniform atomization. Under conditions of continuous use, the actual atomization air pressure and flow volume should remain nearly constant and, ordinarily, should not deviate from the set value by more than 5 percent.

(3) Wire feed rate. The wire feed rate should be adjusted to properly optimize the time spent in the flame. Excessive feed rates may result in inadequate or partial melting of the feedstock and may produce very rough deposited coatings. Too slow a feed rate may cause the wire to be overoxidized and will produce coatings of poor quality. Under conditions of continuous use, the actual wire feed rate should remain nearly constant and, ordinarily, should not deviate from the set value by more than 5 percent.

(4) Powder feed rate. The powder feed rate should be adjusted to properly optimize the time spent in the flame. Excessive feed rates may result in inadequate or partial melting of the powder and may produce very rough deposited coatings. Too slow a feed rate may cause the powder to be overoxidized and will produce coatings of poor quality. Under conditions of continuous use, the actual powder feed rate should remain nearly constant and, ordinarily, should not deviate from the set value by more than 10 percent. The powder feedstock is typically

represented by a range of particle sizes. The various sized powder particles should be consumed at nearly the same rate and should not undergo a size segregation in the hopper.

(5) Air cap selection. A choice of air caps is available for powder flame spray equipment but not for wire flame spray. Wire flame spray is limited to a relatively small-diameter round spray pattern. Air caps for powder flame spray include fan (oval) and round spray patterns.

b. Arc spray.

(1) Power. In general, the higher the power output of the direct current power supply, the greater the possible production rate of the unit. Under conditions of continuous use, the actual current output should remain nearly constant and, ordinarily, should not deviate from the set value by more than 5 percent. Power supplies that are adequately sealed may be operated in dusty atmospheres and do not need to be isolated from the thermal spray operation. DC power supplies rated as high as 600 A are common. A lightweight power supply mounted on pneumatic tires will have added portability. The adhesion and deposit efficiency of the thermal spray coating is dependent on the power. There is an optimal amperage for each coating material that may further depend on wire diameter and the equipment model.

(2) Voltage. The voltage should be adjustable to accommodate different wire materials. Voltage should be set to the lowest level consistent with good arc stability. This will provide smooth dense coatings with superior deposit efficiency. Higher voltages increase thermal spray particle sizes and produce rougher coatings with lower densities. Under conditions of continuous use, the actual voltage should remain nearly constant and, ordinarily, should not deviate from the set value by more than 5 percent.

(3) Wire feed rate. The wire feed mechanism should be designed for automatic alignment. Manual alignment of the wires is time consuming and inexact. The wire feed mechanism must be capable of providing wire at a rate commensurate with the power consumption of the unit. Under conditions of continuous use, the actual wire feed rate should remain nearly constant and, ordinarily, should not deviate from the set value by more than 5 percent.

(4) Atomization air pressure. Under conditions of continuous use, the actual atomization air pressure and flow volume should remain nearly constant and, ordinarily, should not deviate from the set value by more than 5 percent. Lower atomization air pressures will produce rougher coatings with lower densities.

(5) Air cap selection. A choice of air caps is available for arc spray equipment. Air caps for arc spray include fan (oval) and round spray patterns. Some air caps are adjustable.

(6) Cable length. Most manufacturers offer optional cable packages that allow operation up to 30 m (100 ft) from the power supply. Longer cables provide added flexibility for thermal spraying in the field.

(7) Wire tips. Wire tips that hold and align the wires as they enter the arc zone are subject to wear. Properly designed equipment will allow cooler operating temperatures which will prolong tip life and reduce maintenance time. Easy-to-change tips are also beneficial.

(8) Arc shorting control. Arc shorting is a phenomenon wherein the arc shorts and must be restarted. Shorting sometimes requires that the wire ends be manually clipped before the arc is restruck. This operation can be very time consuming. Sometimes during arc shorting, lumps of unmelted wire are sheared off and deposited on the substrate resulting in poor coating quality. An added feature available on some arc spray equipment will control arc shorting.

7-5. Coverage of Thermal Spray Coatings

Nominal values for deposit efficiency, spray rates, and coverage of thermal spray coatings are available in the literature and from manufacturers. Actual rates may vary widely, depending on the actual equipment used, spraying parameters, operator experience, complexity of the item being coated, and whether the work is performed in the shop or in the field. The deposit efficiency is defined as the percentage of material sprayed that is actually deposited on a large flat surface. Table 7-2 presents deposit efficiencies for different thermal spray materials and processes. Table 7-3 shows the amount of material required to coat a unit area with a given thickness of coating. Table 7-4 lists spray rates (kg/hr) for the different processes and materials. Table 7-5 presents the equivalent coverage rates. This information is presented as a general guide and should not be used for cost estimating.

Table 7-2
Deposit Efficiency of Thermal Spray Processes

Material	Wire Flame Spray percent	Powder Flame Spray percent	Arc Spray percent
Zinc	65-70	85-90	60-65
Aluminum	80-85	85-90	70-75
85-15 Zn-Al	85-90	N/A	70-75

Table 7-3
Weight of Material Required for Coating a Given Area (kg/m²/μm (lb/ft²/mil))

Material	Wire Flame Spray	Powder Flame Spray	Arc Spray
Zinc	0.0098 (0.050)	0.0076 (0.039)	0.0110 (0.056)
Aluminum	0.0027 (0.014)	0.0027 (0.014)	0.0029 (0.015)
85-15 Zn-Al	0.0070 (0.036)	N/A	0.0093 (0.048)

Table 7-4
Spray Rates (kg/hr (lb/hr))

Material	Wire Flame 2.4-mm (3/32-in.) wire	Wire Flame 3.2-mm (1/8-in.) wire	Wire Flame 4.8-mm (3/16-in.) wire	Powder Flame Spray	Arc Spray (per 100 amps)
Zinc	9.1 (20.0)	20 (44.0)	30 (66.1)	14 (30.8)	18 (39.6)
Aluminum	2.5 (5.5)	5.4 (11.9)	7.3 (16.1)	6.8 (15.0)	2.7 (5.9)
85-15 Zn-Al	8.2 (18.0)	18 (39.6)	26 (57.3)	N/A	16 (35.2)

Table 7-5
Coverage Rates (m²/hr/μm (ft²/hr/mil))

Material	Wire Flame 2.4-mm (3/32-in.) wire	Wire Flame 3.2-mm (1/8-in.) wire	Wire Flame 4.8-mm (3/16-in.) wire	Powder Flame Spray	Arc Spray (per 100 amps)
Zinc	944 (2190)	2120 (4930)	3070 (7130)	1960 (4550)	1100 (2560)
Aluminum	873 (2030)	1890 (4390)	2530 (5880)	2600 (6040)	826 (1920)
85-15 Zn-Al	1180 (2740)	2620 (6090)	3800 (8830)	N/A	968 (2250)

7-6. Sequence of Thermal Spray Application

Surfaces that will be thermal sprayed must be clean before application of the thermal spray coating. Cleaning, thermal spray application, and sealing should be scheduled so that dust, overspray, and other contaminants from these operations are not deposited on surfaces ready for thermal spray coating or sealing. Surfaces that will not be thermal sprayed should be protected from the effects of blast cleaning and thermal spray application. Special care should be taken to prevent entry of abrasive and thermal spray dusts into sensitive machinery and electrical equipment. Painted surfaces adjacent to surfaces receiving thermal spray coatings should be adequately protected from damage by molten thermal spray particles. Thermal spray coatings should not be applied closer than 2 cm (3/4 in.) to surfaces that will be welded. Surfaces that have been cleaned for thermal spray should be sprayed as soon as practicable. The first thermal spray coat should be applied before the appearance of rust bloom on the surface or within 4 hr of blast cleaning, whichever is sooner. In some cases, it may be possible to hold the appearance of the blast cleaned substrate for longer periods of time using a dehumidification system that supplies dry air to the spray enclosure.

Chapter 8
Sealing and Painting of Thermal Spray Coatings

8-1. Introduction

Sealing and painting of the thermal spray coating system is often performed to improve performance or to achieve a desired appearance. Effective sealers have specific characteristics. Different types of sealer materials should be used depending on the exposure environment and the type of thermal spray metal. Recommended sealer systems for use over zinc, aluminum, and 85-15 thermal spray coatings are identified in Tables 5-1 (freshwater immersion), 5-2 (seawater immersion), and 5-3 (atmospheric exposures). Additional information on these materials for sealing and painting thermal spray coatings may be found in CEGS-09965 and EM 1110-2-3400. High-viscosity, thick film coatings should never be applied directly to an unsealed thermal spray coating.

8-2. Purpose of Sealers

Thermal spray coatings have porosities ranging up to 15 percent. Interconnected or through-porosity may extend from the coating surface to the substrate. Through-porosity may impair the performance of the thermal spray coating. Aluminum coatings less than 150 μm (0.006 in.) and zinc coatings less than 225μm (0.009 in.) thick should be sealed for this reason. Sealers are used to fill porosity and improve the overall service life of the thermal spray system. Thermal spray coatings are also self-sealing. Over time, natural corrosion products fill the pores in the coating. Oxidation consumes a relatively minor amount of the metal coating. In some cases, sealing is performed to improve the appearance and cleanability of the thermal spray coated surface. Sealers reduce the retention of dirt and other contaminants by the thermal spray coating. In particular, the sealer may prevent the accumulation of corrosive salts, rain-borne corrosives, and bird droppings.

8-3. Characteristics of Sealers

Sealer materials should be low-viscosity products that flow and are readily absorbed into the pores of the thermal spray coating. Sealers should generally be low-build products that may be applied at low film thicknesses, generally 75 μm (0.003 in.) or less. The resin chemistry of the sealer must be compatible with the thermal spray metal. Some oleoresinous sealers may saponify if applied over zinc metal surfaces because of the alkalinity of the zinc. The selected sealer material must also be compatible with the intermediate and topcoats of paint if used. Sealers must be suitable for the intended service environment. The paint coats should also be applied to relatively low film thicknesses, generally not exceeding 125 μm (0.005 in.).

8-4. Types of Sealers

a. Vinyl. Vinyl-type coatings are well suited to sealing thermal spray coatings. They are compatible with most service environments including sea- and freshwater immersion and marine, industrial, and rural atmospheres. Vinyls are compatible with zinc, aluminum, and 85-15 coatings. They are very low-viscosity materials with low film build characteristics. Vinyl sealers should be applied to a dry film thickness of about 37.5 μm (0.0015 in.). Vinyl sealers are readily topcoated with vinyl paint. Subsequent coats of vinyl should be applied to a dry film thickness of 50 μm (0.002 in.) per coat. The vinyl sealer should be thinned 25 percent by volume with the specified thinner. The approximate viscosity of the sealer should be 20 to 30 sec measured with a No. 4 Ford cup viscometer in accordance with ASTM D1200 "Test Method for Viscosity by Ford Cup Viscometer." Gray (V-766e), white (V-766e), black (V-103c), red (V-106d), and aluminum (V-102e) vinyl finish coats are available.

b. Epoxy. Three types of epoxy sealers are used, coal tar epoxy (C-200A), aluminum epoxy mastic (CID A-A-3127), and epoxy primer/polyurethane topcoat system (CID A-A-3132).

(1) Coal tar epoxy (C-200A). Coal tar epoxy coating may be used as a relatively thick film single coat sealer for use over zinc, aluminum, and 85-15 zinc-aluminum thermal spray coatings applied to penstocks, spiral cases and extensions, draft tube liners, and surge tanks. The coal tar epoxy sealer should be thinned approximately 20 percent by volume and applied in a single coat to a dry film thickness of 100 to 150 μm (0.004 to 0.006 in.). The sealer is applied at a thickness suitable for covering the roughness of the thermal spray coating, providing a smooth surface that minimizes hydraulic friction.

(2) Aluminum pigmented epoxy mastic (CID A-A-3127). The aluminum epoxy mastic sealer is suitable for one coat use over zinc, aluminum, and 85-15 zinc-aluminum thermal spray coatings for use in marine, industrial, and rural atmospheres as well as for use over aluminum and 90-10 aluminum-aluminum oxide in nonskid applications. The aluminum epoxy mastic should be thinned to the maximum extent recommended in the manufacturer's written directions and applied to a dry film thickness of 75 to 125 μm (0.003 to 0.005 in.). This sealer provides an aluminum finish.

(3) Epoxy primer/polyurethane topcoat (CID A-A-3132). The epoxy sealer urethane topcoat system is suitable for use over zinc, aluminum, and 85-15 zinc-aluminum coatings exposed in marine, industrial, and rural atmospheres as well as for use on nonskid aluminum and 90-10 aluminum-aluminum oxide coatings. The epoxy sealer coat should be thinned to the maximum extent recommended in the manufacturer's written directions and applied to a dry film thickness of 75 to 100 μm (0.003 to 0.004 in.). The polyurethane topcoat should be applied to a maximum dry film thickness of 75 μm (0.003 in.). The polyurethane topcoat may be procured in a variety of colors.

c. Oleoresinous. Two types of oleoresinous sealers are used, tung-oil phenolic aluminum (TT-P-38) and vinyl-butyral wash primer/alkyd (SSPC Paint No. 27/ CID A-A-3132).

(1) Tung-oil phenolic aluminum (TT-P-38). The phenolic aluminum sealer is suitable for use over zinc, aluminum, and 85-15 zinc-aluminum thermal spray coatings exposed in marine, industrial, and rural atmospheres. The sealer should be thinned 15 percent by volume and applied to a dry film thickness of 37.5 μm (0.0015 in.). A second coat of the phenolic aluminum should be applied to the dried sealer to a dry film thickness of approximately 50 μm (0.002 in.). This sealer system produces an aluminum finish.

(2) Vinyl-butyral wash primer/alkyd (SSPC Paint No. 27/ CID A-A-2962). The wash primer-alkyd sealer system is suitable for use over zinc, aluminum, and 85-15 zinc-aluminum thermal spray coatings exposed in marine, industrial, and rural atmospheres. The wash primer coat sealer should be thinned according to the manufacturer's instructions and applied to an approximate dry film thickness of 12.5 μm (0.0005 in.). The commercial alkyd sealer coat should be applied over the dried wash primer coat at a dry film thickness of 50 to 75 μm (0.002 to 0.003 in.). This sealer system is available in a variety of colors and with a gloss or semigloss finish.

d. High temperature.

(1) Aluminum silicone (TT-P-28). The aluminum silicone sealer is suitable for use over thermal spray aluminum and 85-15 zinc-aluminum coatings used for high-temperature applications. The sealer should be thinned 15 percent by volume and applied to a dry film thickness of 25 to 37.5 μm (0.001 to 0.0015 in.). The dried sealer should be topcoated with a second coat of aluminum silicone paint to a dry film thickness of 37.5 to 50 μm (0.0015 to 0.002 in.). This sealer system provides an aluminum finish.

(2) Silicone alkyd. The silicone alkyd sealer is suitable for use over thermal spray aluminum and 85-15 zinc-aluminum coatings used for high-temperature applications. The sealer should be thinned 15 percent by volume and applied to a dry film thickness of 37.5 to 50 μm (0.0015 to 0.002 in.). The dried sealer should be topcoated with a second coat of silicone alkyd paint to a dry film thickness of 37.5 to 50 μm (0.0015 to 0.002 in.). This sealer system is available in a variety of colors.

8-5. Sealing and Painting

In general, surface preparation, thermal spray application, and sealing of a given area should be accomplished in one continuous work period of not longer than 16 hr. Subsequent paint coats should be applied in accordance with the requirements of the painting schedule. Surfaces to be sealed should first be blown down with clean, dry compressed air to remove dust. The thermal sprayed surfaces should be sealed before visible oxidation of the thermal spray coating occurs. Sealers should be applied by conventional or airless spray, except that vinyl-type sealers must only be applied by conventional spray. Spray application ensures the degree of control necessary to achieve thin, uniformly thick coatings. Thin sealer-topcoat systems are preferred to thicker films that may retain moisture and reduce the overall coating system life.

Chapter 9
Thermal Spray Coating Inspection and Testing

9-1. Introduction

This chapter discusses the importance of conducting and documenting various quality control and quality assurance procedures for surface preparation, thermal spraying, and sealing operations. Each procedure is critical in ensuring that a quality coating system is applied.

9-2. Reference Samples and the Thermal Spray Job Reference Standard

a. Reference samples. Reference samples of each material used on a thermal spray job should be collected, including clean, unused abrasive blast media, thermal spray wire, sealer, and paint. Samples may be used to evaluate the conformance of materials to any applicable specifications. A 1-kg (2.2-lb.) sample of blast media should be collected at the start of the job. The sample may be used to verify the cleanliness, media type, and particle size distribution of the virgin blast media. A 30-cm (12-in.) sample of each lot of thermal spray wire should be collected. The wire sample may be used to confirm that the manufactured wire conforms to the size and compositional requirements of the contract. For powdered materials, a 1-kg (2.2-lb.) sample should be collected. One-liter (1-quart) samples of all sealers and paints should be collected for compliance testing.

b. Thermal spray job reference standard. A thermal spray job reference standard (JRS) should be prepared. The JRS may be used at the initiation of a thermal spray contract to qualify the surface preparation, thermal spray application, and sealing processes. The JRS may also serve as a standard of quality in case of dispute.

(1) Preparing the JRS. The JRS should be prepared prior to the onset of production work. To prepare the JRS, a steel plate of the same alloy and thickness, measuring 60 x 60 cm (2 x 2 ft), should be solvent and abrasive-blast cleaned in accordance with the requirements of the contract. The abrasive blast equipment and media used for the JRS should be the same as those that will be used on the job. One-fourth of the JRS plate should be masked with sheet metal, and the thermal spray coating should be applied to the unmasked portion of the plate. The thermal spray coating should be applied using the same equipment and spray parameters proposed for use on the job. The gun should be operated in a manner substantially the same as will be used on the job. The approximate traverse speed and standoff distance during spraying should be measured and recorded. Once the JRS is qualified, the operating parameters should not be altered by the contractor, except as necessitated by the requirements of the job. Two-thirds of the thermal spray coated portion of the JRS should be sealed in accordance with the requirements of the contract. One-half of the sealed area should be painted in accordance with the contract if applicable. The sealer and paint should be applied using the same paint spray equipment that will be used for production. The prepared JRS should be preserved and protected in such a manner that it remains dry and free of contaminants for the duration of the contract. The preserved JRS should then be archived for future reference in the event of a dispute or premature coating failure. Figure 9-1 depicts a representative JRS.

(2) Evaluating the JRS. The surface cleanliness, blast profile shape and depth, thermal spray appearance, thickness, and adhesion, and sealer and paint thicknesses should be determined in accordance with the contract requirements and recorded. Paragraphs 9-10, 9-11, and 9-14 through 9-17 provide additional details on performing these evaluations. The JRS and the measured values may be used as a visual reference or job standard for surface preparation, thermal spray coating, sealing, and painting.

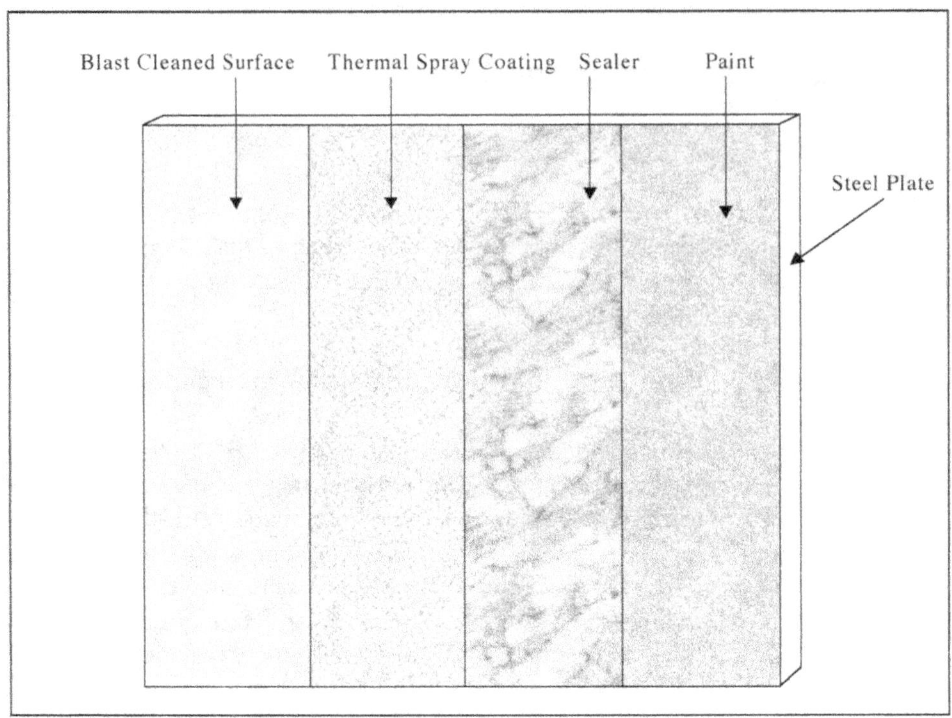

Figure 9-1. Representative job reference standard

9-3. Presurface Preparation Inspection

Prior to abrasive blasting, the substrate should be inspected for the presence of contaminants, including grease and oil, weld flux and spatter, heat-affected zones, pitting, sharp edges, and soluble salts.

(1) Grease and oil. Painted surfaces and newly fabricated steel should be visibly inspected for the presence of organic contaminants such as grease and oil as required by the project specification. A black light may be used to increase the sensitivity of the evaluation, as grease and oil deposits will fluoresce under the light. Solvent evaporation and heat tests may be used to detect thin films of oil contamination on steel surfaces.

(2) Weld flux and spatter. A visual inspection for the presence of weld flux and spatter should be performed as required by the project specification. Weld flux should be removed prior to abrasive blast cleaning using a suitable SSPC-SP 1 "Solvent Cleaning" method. Weld spatter may be removed either before or after abrasive blasting using suitable impact or grinding tools. Areas that are power-tool cleaned of weld spatter should be abrasive blast cleaned.

(3) Heat affected zones. Heat affected zones should be identified and marked prior to abrasive blasting as required by the specification. The demarcated areas should be ground using power tools prior to abrasive blast cleaning.

(4) Pitting. Deep pits or pitted areas should be identified and marked prior to abrasive blast cleaning as required by the specification. The demarcated areas should be ground using power tools prior to abrasive blast cleaning.

(5) Sharp edges. Sharp edges should be identified and marked prior to abrasive blasting as required by the specification. The demarcated edges should be prepared by grinding to a minimum radius of 3 mm (1/8 in.) prior to blast cleaning.

(6) Soluble salts. When soluble salt contamination is suspected, the contract documents should specify a method of retrieving and measuring the salt levels as well as acceptable levels of cleanliness. Salt contamination is prevalent on structures exposed in marine environments and on structures such as parking decks and bridges exposed to deicing salts. Common methods for retrieving soluble salts from the substrate include cell retrieval methods and swabbing or washing methods. Various methods are available for assessing the quantity of salts retrieved, including conductivity, commercially available colorimetric kits, and titration. The rate of salt retrieval is dependent on the retrieval method. The retrieval and quantitative methods should be agreed upon in advance. For additional information on testing for soluble salts, refer to section 9-11, "SSPC Technology Update: Field Methods for Retrieval and Analysis of Soluble Salts on Substrates," and SSPC 91-07. Soluble salt levels should be rechecked for compliance with the specification after solvent cleaning and abrasive blasting have been completed.

9-4. Measuring Ambient Conditions Prior to Blasting

An assessment of the atmospheric conditions should be made prior to the commencement of abrasive blast cleaning. The conditions of humidity, dew point, and ambient air temperature should be measured and recorded. Humidity should be determined in accordance with ASTM E337 "Test Method for Measuring Humidity with a Psychrometer (The Measurement of Wet-Bulb and Dry-Bulb Temperatures)." In general, abrasive blasting should not be performed unless the ambient requirements for thermal spray coating and sealing are met. This is because of the rapid sequencing of surface preparation and thermal spraying and the need for a clean, rust-free surface.

9-5. Assessing Compressed Air Cleanliness

The compressed air used for abrasive blasting, thermal spraying, sealing, and painting should be clean and dry. Oil or water in the blasting air supply may contaminate or corrode the surface being cleaned. Oil or water in the thermal spray, sealing, or painting air supply may result in poor coating quality or reduced adhesion. Compressed air cleanliness should be checked in accordance with ASTM D 4285 "Method for Indicating Water or Oil in Compressed Air." The air compressor should be allowed to warm up, and air should be discharged under normal operating conditions to allow accumulated moisture to be purged. An absorbent clean white cloth should be held in the stream of compressed air not more than 60 cm (24 in.) from the point of discharge for a minimum of 1 min. The air should be checked as near as possible to the point of use and always after the position of the in-line oil and water separators. The cloth should then be inspected for moisture or staining. The compressed air source should not be used if there is any oil or water contamination present.

9-6. Determining Abrasive Cleanliness

Abrasive blast media must be free of oil and salt to prevent contamination of the substrate. Recycled steel grit abrasive should comply with requirements of SSPC-AB 2 "Specification for Cleanliness of Recycled

a. Evaluating for salt in abrasives. Most abrasives used to prepare steel substrates for thermal spraying are unlikely to contain appreciable amounts of soluble salts. However, slag abrasives used for strip blasting may sometimes contain measurable quantities of salts. Slag abrasives should be evaluated in accordance with ASTM D 4940 "Test Method for Conductimetric Analysis of Water Soluble Ionic Contamination of Blasting Abrasives."

b. Testing for oil in abrasives. To test for oil in abrasives, a clear glass container should be half filled with unused abrasive, then distilled or deionized water should be added to fill the container. The resulting slurry mixture is stirred or shaken and allowed to settle. The water is then examined for the presence of an oil sheen. If a sheen is present, the media should not be used, and the source of contamination should be determined and corrected.

9-7. Measuring Blast Air Pressure

The contractor should periodically measure and record the air pressure at the blast nozzle. The measurement should be performed at least once per shift and should be performed on each blast nozzle. Measurements should be repeated whenever work conditions are altered such that the pressure may change. Pressures should be checked concurrently with the operation of all blast nozzles. The method employs a hypodermic needle attached to a pressure gauge. The needle is inserted into the blast hose at a 45-deg angle toward and as close to the nozzle as possible. The blast pressure is read directly from the gauge.

9-8. Examining the Blast Nozzle Orifice

The contractor should visually inspect the blast nozzle periodically for wear or other damage. Gauges are available that insert into the end of the nozzle and measure the orifice diameter. Nozzles with visible damage or nozzles that have increased one size should be replaced. Worn nozzles are inefficient and may not produce the desired blast profile. Damaged nozzles may be dangerous.

9-9. Evaluating Surface Profile

a. ASTM D 4417 "Test Methods for Field Measurements of Surface Profile on Blast Cleaned Steel" Method C is the recommended method for measuring the surface profile depth. Recommended surface profiles contained in this manual are based on values obtained using Method C. Methods A and B may provide different measures of the blast profile than Method C.

b. The method employs a replica tape and spring gauge micrometer to measure the surface profile. With the wax paper backing removed, the replica tape is placed face down against the substrate, and a burnishing tool is used to rub the circular cutout until a uniform gray appearance develops. The replica tape thickness (compressible foam plus plastic backing) is then measured using the spring micrometer. The profile is determined by subtracting the thickness of the plastic backing material, 50 μm (0.002 in.), from the measured value. Three readings should be taken within a 100-cm^2 (16-in^2) area, and the surface profile at that location should be reported as the mean value of the readings. The number of measurements per unit area (e.g., 3 per 45 m^2 (500 ft^2)) should be specified in the contract document. Two types of replica tape are available, coarse (20 to 50 μm (0.0008 to 0.002 in.)) and X-coarse (37.5 to 112.5 μm (.0015 to 0.0045 in.)). In most cases, the X-coarse tape will be used to measure profile. It may be possible to measure profiles as high as 150 μm (0.006 in.) using the X-coarse tape.

9-10. Inspecting Surface Cleanliness

After abrasive blasting and prior to thermal spraying the surface should be inspected for cleanliness, including blast cleanliness, soluble salts, grease and oil, and dust.

a. Blast cleanliness. The final appearance of the abrasive cleaned surface should be inspected for conformance with the requirements of SSPC-SP 5. An SP 5 surface is defined as free of all visible oil, grease, dirt, dust, mill scale, rust, paint, oxides, corrosion products, and other foreign matter. The appearance of SP 5 surfaces is dependent on the initial condition of the steel being cleaned. SSPC VIS 1 may be used to interpret the cleanliness of various blast cleaned substrates based on the initial condition of the steel and the type of abrasive used. Initial conditions depicted include: (1) Rust Grade A - steel surface completely covered with adherent mill scale with little or no rust visible, (2) Rust Grade B - steel surface covered with both mill scale and rust, (3) Rust Grade C - steel surface completely covered with rust with little or no pitting, (4) Rust Grade D - steel surface completely covered with rust with visible pitting. The inspector should determine the initial substrate condition or conditions. The final appearance of the surfaces can then be compared with the appropriate photograph. No stains should remain on the SP 5 surface. However, the appearance of the surface may also vary somewhat, depending on the type of steel, presence of roller or other fabrication marks, annealing, welds, and other differences in the original condition of the steel.

b. Soluble salts. Common methods for retrieving soluble salts from the substrate include cell retrieval methods and swabbing or washing methods. Various methods are available for assessing the quantity of salts retrieved, including conductivity, commercially available colorimetric kits, and titration. The rate of salt retrieval is dependent on the retrieval method. The retrieval and quantitative methods should be agreed upon in advance. The recommended procedure employs the Bresle cell (ISO 8502-6) to extract soluble salts from the substrate. Chloride ion concentration is readily measured in the field using titration strips available from Quantab. The test strip analyzes the collected sample and measures chloride ion concentration in parts per million. The unit area concentration of chloride ion is calculated in μ grams per centimeter. The lower detection limit for the Bresle/Quantab method is about 2 $\mu g/cm^2$. SSPC-SP 12/NACE No. 5 describe levels of soluble salt contamination. It is recommended that surfaces cleaned to an SC-2 condition be used for thermal spray coatings. An SC-2 condition is described as having less than 7 $\mu g/cm^2$ of chloride contaminants, less than 10 $\mu g/ cm^2$ of soluble ferrous ion, and less than 17$\mu g/cm^2$ of sulfate contaminants. Most USACE structures have low levels of soluble salt contamination, and, therefore, testing is not usually warranted. Structures that are likely to have soluble salt contamination, including those in marine or severe industrial atmospheres, bridge or other structures exposed to deicing salts, and seawater immersed structures, should be tested. The number of tests per unit area (e.g., 1 per 90 m^2 (1000 ft^2)) should be specified in the contract documents.

c. Grease and oil. Blasted surfaces should be visibly inspected for the presence of grease and oil. A black light (ultraviolet) may be used to increase the sensitivity of the evaluation, as grease and oil deposits will fluoresce under the light. Fluorescence cannot be detected in daylight so a hood or shield must be used to darken the viewing area. Grease or oil contamination is indicated by a yellow or green fluorescence. An absence of fluorescence indicates a clean surface. However, it should be noted that some synthetic oils do not fluoresce. Solvent evaporation and heat tests may also be used to detect thin films of oil contamination. The evaporation test uses a small amount (5 ml) of a residueless, highly volatile solvent, such as acetone, on the surface. The solvent is applied and allowed to evaporate. A visible ring of residue signals the presence of oil or grease contamination. The heat test uses a propane torch to heat the surface to 120 °C (250 °F) to perform a similar visual assessment.

d. Dust. Abrasive blasting, and overspray from painting or metallizing, can leave a deposit of dust on the cleaned substrate. The dust may interfere with adhesion of the thermal spray coating. Residual dust may be detected by applying a strip of clear tape to the substrate. The tape is removed and examined for adherent particles. Alternatively, a clean white cloth may be wrapped around a finger and wiped across the surface. The cloth and substrate are then examined for signs of dust. The preferred method of removing residual dust is by vacuuming. Alternatively, the surface may be blown down with clean, dry compressed air.

9-11. Measuring Ambient Conditions Prior to Thermal Spraying

A second assessment of the atmospheric conditions should be made before thermal spray application begins. The conditions of humidity, dew point, and ambient air temperature should be measured and recorded. Application should not be performed unless the ambient requirements for thermal spraying and sealing are met. This is because of the rapid sequencing of thermal spraying and sealing.

9-12. Bend Testing to Evaluate Equipment Setup

Each day, or every time the thermal spray equipment is to be used, the inspector should record and confirm that the operating parameters are the same as those used to prepare the JRS. The thermal spray applicator should then apply the coating to prepared test panels and conduct the bend test. The bend test is a qualitative test used to confirm that the equipment is in proper working condition. The test consists of bending coated steel panels around a cylindrical mandrel and examining the coating for cracking. If the bend test fails, corrective actions must be taken prior to the application of the thermal spray coating. The results of the bend test should be recorded and the test panels should be labeled and saved.

a. Test panels. The test panels should be a cold rolled steel measuring 7.5 × 15 × 1.25 cm (3 × 6 × 0.050 in.). The panels should be cleaned and blasted in the same fashion as will be used for the job.

b. Application of thermal spray. The thermal spray coating should be applied to five test panels using the identical spray parameters and average specified thickness that will be used on the job. The coating should be applied in a cross hatch pattern using the same number of overlapping spray passes as used to prepare the JRS. The coating thickness should be measured to confirm that it is within the specified range.

c. Conduct bend test. Test panels are bent 180 deg around a steel mandrel of a specified diameter. Thermal spray coating systems 1-Z, 2-Z, 4-Z-A, 5-Z-A, 7-A, and 8-A should be tested using a 12.5-mm- (0.5-in.-) diam mandrel. Systems 3-Z and 6-Z-A should be tested using a 15.6-mm- (0.625-in.-) diam mandrel. Pneumatic and manual mechanical bend test apparatuses may be used to bend the test panels.

d. Examine bend test panels. Test panels should be examined visually without magnification. The bend test is acceptable if the coating shows no cracks or exhibits only minor cracking with no lifting of the coating from the substrate. If the coating cracks and lifts from the substrate, the results of the bend test are unacceptable. Thermal spray coatings should not be applied if the bend test fails, and corrective measures must be taken. Figure 9-2 depicts representative bend test results.

9-13. Measuring the Coating Thickness

The thickness of the thermal spray coating should be evaluated for compliance with the specification. Magnetic film thickness gauges such as those used to measure paint film thickness should be used. Gauges should always be calibrated prior to use. Thickness readings should be made either in a straight line with individual readings taken at 2.5-cm (1-in.) intervals or spaced randomly within a 25-cm^2 (4-in^2) area. Line measurements should be used for large flat areas and area measurements should be used on complex surface geometries and surface transitions such as corners. The average of five readings comprises one thickness measurement. A given number of measurements per unit area (e.g., 5 per 9 m^2 (100 ft^2)) should be specified in the contract documents. Areas of deficient coating thickness should be corrected before sealing begins.

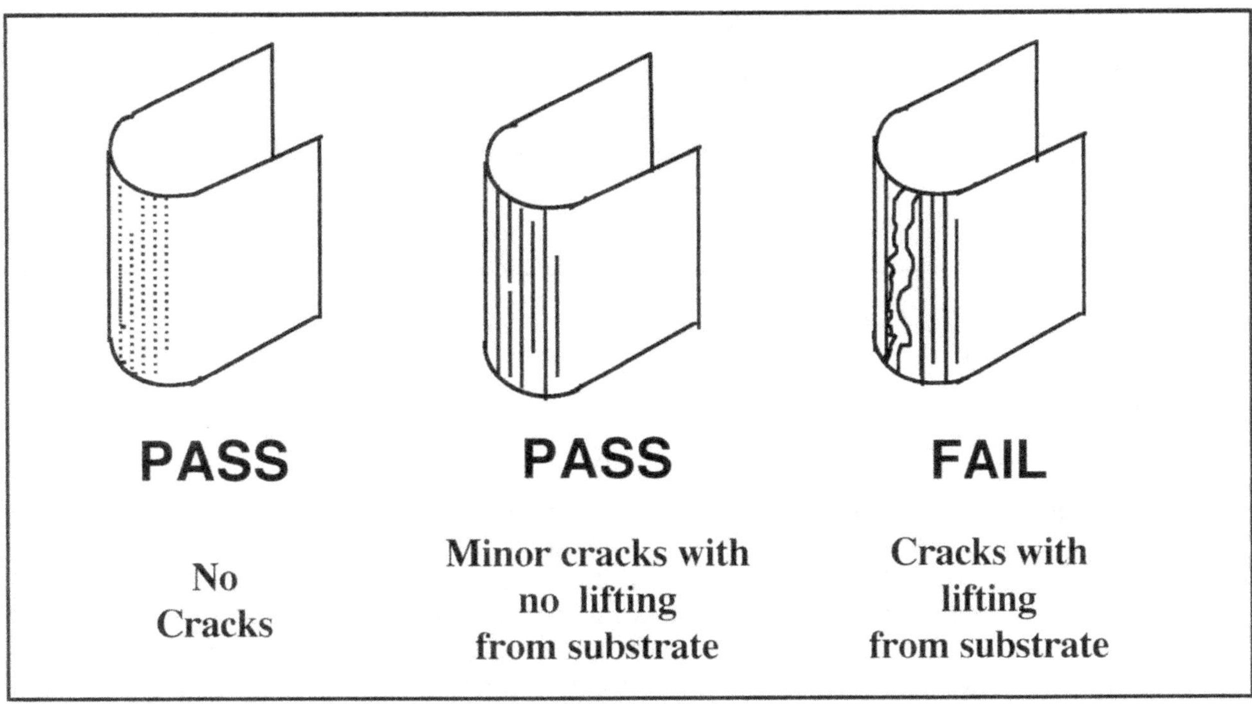

Figure 9-2. Representative pass and fail conditions for qualitative bend test

9-14. Inspecting the Appearance of the Applied Coating

The appearance of the applied thermal spray coating should be inspected for obvious defects related to poor thermal spray applicator technique or equipment problems. The coating should be inspected for the presence of blisters, cracks, chips or loosely adhering particles, oil, pits exposing the substrate, and nodules. A very rough coating may signal that the coating was not applied with the gun perpendicular to the surface or with too high a standoff distance. Coatings that appear oxidized or powdery should be evaluated by light scraping. If scraping fails to produce a silvery metallic appearance, the coating is defective.

9-15. Adhesion Testing for Quality Control

The adhesion of the thermal spray coating should be evaluated for compliance with the specification in accordance with ASTM D 4541. A self-aligning-type IV tester, described in Annex A4 of ASTM D 4541, should be used. A total of three adhesion tests should be performed in a 100-cm^2 (16-in^2) area, and the average of the three tests should be reported as a single measurement. The number of measurements per unit area (e.g., 1 per 45 m^2 (500 ft^2)) should be specified in the contract documents. Areas of deficient adhesion should be abrasive blasted, and the coating should be reapplied. Additional testing will probably be necessary to determine the extent of the area with poor adhesion. Adhesion testing should be minimized because the test method destroys the coating. Areas damaged by adhesion testing must be repaired by abrasive blasting and reapplication of the metallic coating. Adhesion testing is performed in a small area (100 cm^2 (16 in^2)) to limit the area that must be repaired. As an alternative to testing the adhesion to the failure point, the tests can be interrupted when the minimum specified adhesion value is achieved. This method precludes the need to repair coatings damaged by the test. The adherent pull stubs can then be removed by heating to soften the glue or by firmly striking the side of the stub. Table 9-1

Table 9-1
Typical Adhesion of Field- and Shop-Applied Thermal Spray Coatings Measured by Pull-Off Testing

Thermal Spray Material	Adhesion, kPa (psi)
Zinc	5100 (750)
Aluminum	10,880 (1600)
85-15 Zinc-aluminum	6800 (1000)

lists the recommended adhesion requirements for field- or shop-applied thermal spray coatings of zinc, aluminum, and 85-15 zinc-aluminum.

9-16. Inspecting the Sealer Coating

The dry film thickness of the sealer and paint coats should be evaluated for compliance with the specification in accordance with ASTM D 4138 "Test Methods for Measurement of Dry Film Thickness of Protective Coating Systems by Destructive Means." Test Method A should be used. This method uses a tungsten carbide-tipped instrument to scribe through the sealer and paint leaving a v-shaped cut. A heavy dark colored marking pen is first used to mark the coated surface. The scribing instrument is then drawn across the mark. This process sharply delineates the edges of the scribe. A reticle equipped microscope is used to read the film thickness. A total of three thickness readings should be performed in a 100-cm^2 (16-in.^{2}) area, with the average of the three tests reported as a single measurement. The number of measurements per unit area (e.g., 1 per 45 m^2 (500 ft^2)) should be specified in the contract documents. Areas of deficient thickness should be noted and corrected, if practicable, by adding sealer or paint. Additional testing may be necessary to determine the extent of the area with deficient sealer or paint thickness. Thickness testing should be minimized because the test method destroys the sealer and paint. Areas damaged by adhesion testing must be repaired by touchup with sealer or paint using a brush or spray gun. Thickness testing is performed in a small area (100 cm^2 (16 in.^{2})) to limit the area that must be repaired. The sealer thickness should be checked prior to the application of paint coats, if practical, and the measurement procedure should be repeated for the sealer and paint.

9-17. Frequency of Inspection

The required frequency of inspection procedures should be spelled out in the specification. Inspection can be expensive, and care should be taken not to overspecify inspection procedures. Conversely, inspection has an inherent value that is sometimes intangible. It is difficult to measure the value added by inspection resulting from the conscientious performance of the contract. Thermal spray can be quite sensitive to the quality of surface preparation, thermal spray equipment setup, and application technique. Therefore, it is important to specify an appropriate level of inspection. Table 9-2 presents recommended frequencies for various inspection procedures.

Table 9-2
Recommended Inspection Frequencies for Selected Procedures

Inspection Procedure	Recommended Frequency per Unit Area
Surface profile	3 per 45 m^2 (500 ft^2)
Thermal spray coating thickness	5 per 9 m^2 (100 ft^2)
Thermal spray adhesion	1 per 45 m^2 (500 ft^2)
Sealer thickness	1 per 45 m^2 (500 ft^2)
Paint thickness	1 per 45 m^2 (500 ft^2)
Soluble salts	1 per 90 m^2 (1000 ft^2)

9-18. Documentation

The documentation of inspection activities provides a permanent record of the thermal spray job. Thorough documentation provides a written record of the job in the event of a contract dispute or litigation. Inspection records may also be used to help diagnose a premature coating failure. Future maintenance activities may be simplified by the existence of complete inspection records. As a minimum, at least one full-time inspector should be used on all thermal spray jobs to ensure adequate inspection and documentation. Inspection should be performed by a USACE inspector or a qualified third party inspector from a reputable firm. As a minimum, the inspector should perform and document the inspection procedures described in this chapter. Sample documentation forms for industrial coating activities are widely available through various sources.

Chapter 10
Thermal Spray Applicator and Equipment Qualification

10-1. Introduction

The use of qualified thermal spray equipment and applicators helps to ensure that a quality thermal spray coating will be achieved. The qualified equipment should perform smoothly and consistently to produce dense, firmly adherent coatings without visible defects. The qualified applicator should be knowledgeable in the setup and operation of the equipment and should be able to apply a sample coating that meets the requirements of the contract.

10-2. Equipment Qualification Procedure

Each type and source of thermal spray equipment should be qualified prior to use. The equipment should conform to the following requirements related to uniformity of operation, coating appearance, and coating adhesion. Equipment should be qualified using the type and size of wire to be used on the job. The operating parameters should be those selected by the contractor for use on the job. Equipment manufacturers may also qualify their equipment for use with specific feedstocks and operating parameters. Such qualified equipment should be accepted as prequalified, assuming the contractor proposes to operate the equipment in the same manner used for the qualification tests.

a. Uniformity of operation.

(1) Wire and powder flame spray equipment.

(a) Oxygen and fuel gas flow rates. Under conditions of continuous use, the actual oxygen and fuel gas flow rates and pressures should remain nearly constant and should not deviate from the set values by more than 5 percent during a 15-min period.

(b) Atomization air pressure. Compressed air should be free of oil and water. Under conditions of continuous use, the actual atomization air pressure and flow volume should remain nearly constant and should not deviate from the set value by more than 5 percent during a 15-min period.

(c) Wire feed rate. Under conditions of continuous use, the actual wire feed rate should remain nearly constant and should not deviate from the set value by more than 5 percent during a 15-min period.

(d) Powder feed rate. Under conditions of continuous use, the actual powder feed rate should remain nearly constant and ordinarily should not deviate from the set value by more than 10 percent during a 15-min period.

(e) Continuous operation. When operated continuously for 15-min, the equipment should not sputter, pop, or stop operating.

(2) Arc spray equipment.

(a) Power. Under conditions of continuous use, the actual current output should remain nearly constant and should not deviate from the set value by more than 5 percent during a 15-min period.

(b) Voltage. Under conditions of continuous use, the actual voltage should remain nearly constant and should not deviate from the set value by more than 5 percent during a 15-min period.

(c) Wire feed rate. The wire feed mechanism should be designed for automatic alignment. Under conditions of continuous use, the actual wire feed rate should remain nearly constant and should not deviate from the set value by more than 5 percent during a 15-min period.

(d) Atomization air pressure. Under conditions of continuous use, the actual atomization air pressure and flow volume should remain nearly constant and should not deviate from the set value by more than 5 percent during a 15-min period.

(e) Continuous operation. When operated continuously for 15-min, the equipment should not sputter, pop, or stop operating.

(f) On/off operation. The equipment should be capable of continuous start and stop operation for a minimum of 15 cycles consisting of 10 sec on and 5 sec off without fusing, sputtering, or deposition of nodules.

b. Coating appearance. The applied coating should be uniform and free of blisters, cracks, loosely adherent particles, nodules, and powdery deposits.

c. Coating adhesion. A 30- × 30- × 1.25-cm (12- × 12- × 0.5-in.) flat steel plate should be cleaned and prepared in accordance with SSPC-SP 1 and SSPC-SP 5. No. 36 aluminum oxide grit should be used to produce an angular blast profile of 75 ± 5 μm (0.003 ± 0.0002 in.). The blast profile should be measured and recorded using replica tape in accordance with ASTM D 4417. The coating (400 ± 50 μm (0.016 ± 0.002 in.) of 85-15 zinc-aluminum alloy, or 400 ± 50 μm (0.016 ± 0.002 in.) of zinc, or 250 ± 50 μm (0.010 ± 0.002 in.) of aluminum) should be applied in not less than two half-lapped passes applied at right angles to each other. The adhesion should be tested in accordance with ASTM D 4541 using a self-aligning type IV adhesion tester as described in Annex A4 of the method. Scarified aluminum pull stubs should be attached to the thermal spray coating using a two-component epoxy adhesive. The adhesive strength of the coating should be measured and recorded at five randomly selected locations. The average adhesion should not be less than 6800 kPa (1000 psi), 10,880 kPa (1600 psi), and 5100 kPa (750 psi) for 85-15 zinc-aluminum alloy, aluminum, and zinc coatings, respectively. If the test fails, it should be repeated using a new test plate. If the adhesion fails on the second plate, the equipment should be deemed unacceptable.

10-3. Applicator Qualification Procedure

a. Equipment setup and operation. The qualified applicator should be able to demonstrate a working knowledge of the application equipment to be used on the job by proper setup and operation of the equipment. The applicator should prepare a 30- × 30- × 1.25-cm (12- × 12- × 0.5-in.) flat steel plate cleaned in accordance with SSPC-SP 1 and SSPC-SP 5. Aluminum oxide or steel grit should be used to produce an angular blast profile of 3.0 ± 0.2 mils. The blast profile should be measured and recorded using replica tape in accordance with ASTM D 4417. The applicator should apply the coating (400 ± 50 μm (0.016 ± 0.002 in.) of 85-15 zinc-aluminum alloy, or 400 ± 50 μm (0.016 ± 0.002 in.) of zinc, or 250 ± 50 μm (0.010 ± 0.002 in.) of aluminum) using the proper spray technique.

b. Coating appearance. The qualified applicator will have applied a coating that is uniform and free of blisters, cracks, loosely adherent particles, nodules, or powdery deposits.

c. Coating adhesion. The qualified applicator should be able to apply a firmly adherent coating that meets the adhesion requirements of the contract. The thermal spray coating adhesion should be tested in accordance with ASTM D 4541 using a self-aligning type IV adhesion tester as described in Annex A4 of the method. Scarified aluminum pull stubs should be attached to the thermal spray coating using a two-component epoxy adhesive. The adhesive strength of the coating should be measured and recorded at five randomly selected locations. The average adhesion should not be less than 6800 kPa (1000 psi), 10,880 kPa (1600 psi), and 5100 kPa (750 psi) for 85-15 zinc-aluminum alloy, aluminum, and zinc coatings, respectively.

Chapter 11
Maintenance of Thermal Spray Coatings

11-1. Introduction

Thermal spray coatings are very durable and long-lived. However, thermal spray coatings may suffer damage and will eventually degrade in service and, therefore, may require maintenance or repair. The American National Standards Institute (ANSI) and the American Welding Society (AWS) have published a standard for repair and maintenance of thermal spray coatings.

11-2. Assessing the Condition of Thermal Spray Coatings

The first step in the repair of thermal spray coatings is an assessment of the type of thermal spray system and the nature of the damage or wear. The type of thermal spray material originally applied must be determined. If a sealer and paint topcoat are present, they must also be identified. Historic records are the easiest means of making these determinations. Chemical analyses may also be used, but this can be time consuming and expensive. The experienced observer may also be able to distinguish between the various types of thermal spray materials. The type and extent of degradation should be observed and noted. A worn coating is indicated by general or localized thickness reductions. Coating oxidation will be evident by the presence of a powdery residue on the coating surface. The presence of rust and bare steel should be noted. Cracked, blistered, and delaminated areas should be identified. The extent and location of the damage or wear should be determined.

11-3. Repair Procedures

The thermal spray coating repair procedures used depend on the type and extent of degradation and the presence or absence of sealer and paint topcoat. ANSI/AWS C2.18-93 "Guide for the Protection of Steel with Thermal Sprayed Coatings of Aluminum and Zinc and Their Alloys and Composites" addresses the maintenance and repair of thermal spray coatings. This section summarizes the types of repair procedures available. Paragraph 11-4 discusses when and in what combinations these procedures should be used.

a. Solvent clean. Grease and oil should be removed by solvent cleaning using ASTM D 3734 "High-Flash Aromatic Naphthas, Type I." The solvent may be applied by wiping, brushing, or spraying.

b. Scrape with flexible blade. A small (25-mm (1-in.)) flexible blade scaper is used to remove loose paint and thermal spray coating near the damaged or worn coating.

c. Scrape with hard blade. A hard blade scraper is worked underneath the coating to remove loosely adherent paint and thermal spray.

d. Brush clean by hand. A stiff stainless steel or bristle brush is used to remove loose debris by hand brushing.

e. Abrasive brush blast clean. Fine mesh (30 - 60) angular iron oxide grit or aluminum oxide is used to remove loose paint and thermal spray coating. The blasting must be done using a relatively low nozzle pressure (about 340 kPa (50 psi)) so that only loose materials are removed and an anchor profile is created.

f. Power tool clean. Hand-held power tools including disc sanders and rotary brushes should be used with light pressure to clean and roughen the surface. Care should be taken not to polish the surface.

g. Abrasive blast clean to SSPC-SP 10/NACE #2 "Near White Metal." The surface is cleaned and profiled using abrasive blast cleaning equipment. The blast nozzle must be worked perpendicular to the surface to prevent removal of intact thermal spray adjacent to the cleanup area.

h. Feather edges. A 50- to 75-mm (2- to 3-in.) feathered or tapered border is created around the repaired areas.

i. Light abrasion. The cleaned and feathered areas are lightly abraded using sandpaper to improve the adhesion of subsequent sealer and paint coats.

j. Apply thermal spray coating. The thermal spray repair metal should be the same as that originally applied. Flame sprayed coatings should not be topcoated using arc spray equipment. Arc sprayed coatings may be repaired using either arc or flame spray.

k. Seal and topcoat. Sealer and topcoat materials are applied as the final step of the repair sequence.

11-4. Repair Sequences

a. Increasing thermal spray coating thickness. Unsealed thermal spray coatings that are worn thin or that were applied to less than the specified thickness may be repaired by preparing the surface and applying more metal. If the coating was recently applied, it may be possible to simply apply additional coating directly onto the original coating. If the coating is oxidized, the abrasive brush blast procedure should be used prior to application of additional thermal spray coating material.

b. Repair of small damaged areas with steel substrate not exposed. Small damaged areas (< 0.1 m^2 (<1 ft^2)) where the steel substrate is not exposed are repaired by solvent cleaning, scraping with a flexible blade tool, wire brushing, edge feathering, lightly sanding to abrade the cleaned areas, and sealing and painting.

c. Repair of large damaged areas with steel substrate not exposed. Large damaged areas (> 0.1 m^2 (>1 ft^2)) where the steel substrate is not exposed are repaired by solvent cleaning, abrasive brush blasting, edge feathering, and sealing and painting.

d. Repair of thermal spray coatings with steel substrate exposed. Either of two procedures may be used to repair thermal spray coatings damaged to the extent that the steel substrate is exposed. One method uses a rapid "paint only" repair procedure and the other utilizes thermal spray coating plus sealer and paint coats.

(1) The rapid paint repair procedure includes solvent cleaning, scraping with a hard blade tool, power tool cleaning, edge feathering, and sealing and painting.

(2) The thermal spray repair procedure includes solvent cleaning, scraping with a hard blade tool, abrasive blast cleaning to near white metal, edge feathering, thermal spray coating application, and sealing and painting.

Chapter 12
Safety

12-1. Introduction

Surface preparation, thermal spray, and sealing and painting operations expose workers to numerous potential health and safety hazards. The industrial protective coatings industry is considered a high-risk occupation. Common health and safety hazards associated with the industry include (a) electric shock, (b) flammable and explosive solvents, gases, and dusts, (c) confined space entry, (d) fall hazards, (e) exposure to high intensity noise and ultraviolet light and toxic materials, and (f) high-velocity particle impingement. This chapter will discuss the importance of worker safety and specifically addresses safe working practices and contract requirements, including safety-related plans and submittals. Chapter 10 of EM 1110-2-3400 provides additional, more detailed information related to surface preparation and painting.

12-2. Safe Surface Preparation Procedures

a. Hoses and nozzles. Hose and hose connections that do not allow electrostatic discharge should be used. Hose couplings and nozzles should be designed to prevent accidental disengagement. A deadman control device that automatically stops the flow of air and abrasive when the hose is dropped should be used. Hoses and fittings used for abrasive blasting should be inspected frequently to ensure the timely replacement of worn parts and equipment.

b. Abrasive blasting respirator. Abrasive blasting operators should wear an abrasive blasting respirator, which consists of a continuous-flow airline respirator constructed so that it covers the worker's head, neck, and shoulders from rebounding abrasive. Respiratory equipment should be approved by the National Institute for Occupational Safety and Health and/or Mine Safety and Health Administration (NIOSH/MSHA). Compressed air should meet at least the requirements of the specification for Type 1 Grade D breathing air as described in CGA G 7.1 "Commodity Specification for Air."

c. Personal protective equipment. Blasting operators should wear heavy canvas or leather gloves and an apron or coveralls. Safety shoes should be worn to protect against foot injury. Hearing protection should be used during all blasting operations.

d. Cleaning with compressed air. Cleaning with compressed air should be restricted to systems where the air pressure has been reduced to 204 kPa (30 psi) or less. Cleaning operators should wear safety goggles or face shield, hearing protection, and appropriate body covering. Compressed air or pressurized gas should never be pointed at anyone.

e. Cleaning with solvents. The material safety data sheet (MSDS) should be consulted for specific solvent information and procedures in addition to those listed here. Flammable liquids with a closed-cup test flash point below 38 °C (100 °F) should not be used for cleaning purposes. Sources of ignition should not be permitted in the vicinity of solvent cleaning if there is any indication of combustible gas or vapor present. Measurements should be made to ensure that solvent vapors are not present during thermal spray operations, especially in confined spaces. Representative air samples should be collected from the breathing zone of workers involved in the cleaning process to determine the specific solvent vapor concentrations. Worker exposures should be controlled to levels below the Occupational Safety and Health Standards (OSHA) Permissible Exposure Limit as indicated in CFR 29 Part 1910, Section 1000.

f. Additional information. EM 1110-2-3400, Chapters 10 and 11, offers additional information on safety during surface preparation.

12-3. Safe Thermal Spray Procedures

a. General. Airborne metal dusts, finely divided solids, or other particulate accumulations should be treated as explosive materials. Proper ventilation, good housekeeping, and safe work practices should be maintained to prevent the possibility of fire and explosion. Thermal spray equipment should never be pointed at a person or flammable material. Thermal spraying should not be performed in areas where paper, wood, oily rags, or cleaning solvents are present. Conductive safety shoes should be worn in any work area where explosion is a concern. During thermal spray operations, including the preparation and finishing processes, employees should wear protective coveralls or aprons, hand protection, eye protection, ear protection, and respiratory protection.

b. Eye and skin protection. All thermal spray processes introduce particulates into the air that may damage the eyes or skin. Eye and face protection must be worn to protect against particulate impingement. High intensity visible, infrared, and ultraviolet light may also cause eye and skin damage. Flame resistant clothing should be worn to protect the skin. Hoods or face shields conforming to ANSI Z87.1 and ANSI Z89.1 with filter lenses should be worn to protect the face and eyes. Various shades of lens filters are recommended based on the type of thermal spray process being used. For wire flame spray, lens shades 2-4 should be used. For powder flame spray, lens shades 3-6 should be used. For wire arc and powder plasma spray, lens shades 9-12 should be used. Shades 3-6 can be used for wire arc spray if the gun is equipped with an arc shield. The shield encloses the arc and reduces the operator's exposure to high intensity light. Other workers in the vicinity of the thermal spray applicator should also use proper eye protection. Table 12-1 summarizes the recommended lens shades for thermal spraying.

Table 12-1
Recommended Lens Shades for Thermal Spraying

Thermal-Spray Process	Shade
Wire flame spraying	2-4
Powder flame spraying	3-6
Wire arc and powder plasma spraying	9-12
Wire arc if gun is equipped with arc shield	3-6

c. Hearing protection. Thermal spray produces very high intensity noise that will cause permanent hearing loss. Thermal spray operators and other workers in the vicinity of the thermal spray operation should wear hearing protection at all times. Protection against the effects of noise exposure should be provided in accordance with the requirements of EM 385-1-1, Section 5, "Personal Protective and Safety Equipment," Subsection 05.C, "Hearing Protection and Noise Control," and CFR 29 Part 1910, Section 95. Insert earplugs should be used in conjunction with wire or powder flame spray. Insert earplugs should be worn as a minimum for wire arc spray. Insert earplugs and earmuffs are recommended for use with wire arc and plasma spray. Table 12-2 lists the minimum recommended hearing protection devices for various thermal spray application methods.

d. Respiratory protection. Thermal spray produces toxic dusts and fumes. Thermal spray operators and those in the general vicinity of the operation should wear appropriate respirators. Overexposure to zinc fume is known to produce flu-like symptoms, often called metal fume fever.

Table 12-2
Typical Noise Levels and Recommended Hearing Protection for Thermal Spray Processes

Thermal Spray Process	Noise Level dBA	Minimum Recommended Protection
Wire flame spraying	114	Earplugs
Powder flame spraying	90-100	Earplugs
Wire arc	111-116	Earplugs and earmuffs
Plasma spraying	128-131	Earplugs and earmuffs

e. Electrical shock prevention. Wire arc spray presents an electrical shock hazard. The manufacturer's safe operating procedures should always be followed. Ground protection for equipment and cords should be present and in good condition. Electrical outlets in use should have Ground Fault Circuit Interrupters (GFCI) in addition to appropriate over-current protection. Electrical circuit grounds and GFCI should be tested before work begins. Switches and receptacles should have proper covers. Damaged cords and equipment should be repaired or replaced. Circuit breaker boxes should be kept closed. Cords should be approved for wet or damp locations. The cords should be hard usage or extra hard usage as specified in NFPA 70 "National Electrical Code." Cords should not be spliced.

f. Gas cylinder safety. Compressed gas cylinders should be handled in accordance with ANSI Z49.1 and with CGA P-1. Only special oxidation-resistant lubricants should be used with oxygen equipment; grease or oil should not be used. Manifolding and pressure reducing regulators, flow meters, hoses, and hose connections should be installed in accordance with ANSI Z49.1. A protective shield should be used to shield glass tube flow meters from the spray gun. Pressure connecting nuts should be tight, but not overtight. Fittings that cannot be sealed without excessive force, should be replaced. Compressed air for thermal spraying or blasting operations should be used only at pressures recommended by the equipment manufacturers. The airline should be free of oil and moisture. Compressed air, oxygen, or fuel gas should never be used to clean clothing.

g. Flame spray safety. Flame spray equipment should always be maintained and operated according to the manufacturer's instructions. Thermal spray operators should be trained and familiar with their equipment before starting an operation. Valves should be properly sealed and lubricated. Friction lighters, pilot light, or arc ignition methods of lighting flame spray guns should be used. If the flame spray gun backfires, it should be extinguished immediately. Re-ignition of a gun that has backfired or blown out should not be attempted until the cause of the trouble has been determined. Flame spray guns or hoses should not be hung on regulators or cylinder valves. Gas pressure should be released from the hoses after equipment is shut down or when equipment will be left unattended. Lubricating oil should not be allowed to enter the gas mixing chambers when cleaning flame spray guns. Only special oxidation-resistant lubricants should be used on valves or other parts of flame spray guns that are in contact with oxygen or fuel gases.

h. Ventilation. Local exhaust or general ventilation systems should be used to control toxic fumes, gases, or dusts in any operations not performed in the open. Thermal spray should be performed with appropriate respiratory protection and adequate ventilation. Thermal spraying in an enclosed space should be performed with general mechanical ventilation, airline respirators, or local exhaust ventilation sufficient to reduce the fumes to safe limits specified by ACGIH-02. Employee exposures should be controlled to the safe levels recommended by ACGIH-02 or prescribed by CFR 29 Part 1910, whichever is more stringent. Air sampling should be performed before entry to a confined space, during confined-space entry that involves contaminant-generating operations such as flame spray operations, and in areas where ventilation is inadequate to ensure that air contaminants will not accumulate. Engineering controls (enclosures and/or hoods with ducted, mechanical ventilation of sufficient volume to remove contaminants from the work space) are the most-desired methods of preventing job-related illness resulting from breathing air contaminated with harmful dusts, mists, fumes, vapors, or gases.

12-4. Safe Sealing and Painting Procedures

a. Respiratory protection. Sealers and paint coats are typically applied by spray application. Spray application is a high-production rate process that may rapidly introduce very large quantities of toxic solvents into the air. Ventilation should be used to reduce worker exposure to solvents. In most cases, the operator must also use a respirator of the appropriate type. EM 1110-2-3400, paragraphs 10-5, 10-7, and 10-8, provides additional information on respiratory protection.

b. Spray gun operation. Airless spray systems operate at very high pressures. Very high fluid pressures can result in penetration of the skin. Tip guards and trigger locks should be used on all airless spray guns. The operator should never point the spray gun at any part of the body. The pressure remains in the system even after the pump is turned off and can only be relieved by discharging through the gun.

c. Ventilation. Good ventilation is needed when sealing and painting in enclosed areas. Proper ventilation reduces worker exposures to toxic solvents and reduces the risk of explosion. EM 1110-2-3400, paragraphs 10-5 and 10-8, provides additional information on ventilation during painting.

12-5. Safety Plans and Submittals

Thermal spray contracts require the development and submittal of various procedures and plans, including an accident prevention plan, confined space procedures, medical surveillance documentation, a respiratory protection program, an air sampling plan, a worker hazard communication program, a ventilation assessment plan, a qualifications and experience statement, and a safety indoctrination plan.

a. Accident prevention plan. The contractor should prepare a written accident prevention plan that complies with requirements of EM 385-1-1 Appendix A, "Minimum Basic Outline for Accident Prevention Plan." The Accident Prevention Plan should be prepared by a qualified occupational safety and health professional who has a minimum of 3 years experience in safety and industrial hygiene. The accident prevention plan shall address the following requirements as a minimum:

(1) Identification of contractor personnel responsible for accident prevention.

(2) Methods contractor proposes to coordinate the work of subcontractors.

(3) Layout plans for temporary buildings, construction of buildings, use of heavy equipment, and other facilities.

(4) Plans for initial and continued safety training for each of the contractor's and subcontractor's employees.

(5) Plans for traffic control and the marking of hazards to cover waterways, highways and roads, railroads, utilities, and other restricted areas.

(6) Plans for maintaining good housekeeping and safe access and egress at the jobsite.

(7) Plans for fire protection and other emergencies.

(8) Plans for onsite inspections by qualified safety and health personnel. Plans shall include safety inspections, industrial hygiene monitoring if required, records to be kept, and corrective actions to be taken.

(9) Plans for performing Activity Hazard Analysis for each major phase of work. The Activity Hazard Analysis shall include the sequence of work, specific hazards that may be encountered, and control measures to eliminate each hazard.

b. Confined space procedures. The contractor should prepare a written confined-space procedure in compliance with EM 385-1-1, Section 6, "Hazardous Substances Agents, and Environments," Subsection 06.I, "Confined Space," as well as any applicable Federal or local laws. The written procedure should include, but not be limited to, the following requirements:

(1) A description of the methods, equipment, and procedures to test for oxygen content and combustible and toxic atmospheres in confined spaces prior to entry and during work.

(2) Emergency procedures for each type of confined space work, including methods of communication, escape, and rescue.

(3) Air monitoring plans by qualified individuals and a certificate of calibration for all air monitoring equipment.

(4) A plan for training workers in confined-space procedures that should include confined-space hazards, evaluation of confined-space atmospheres, combustible-gas indicator operation, entry procedures, attendant requirements, isolation and lockout, preparation of confined areas, respiratory protection, communication, safety equipment, no smoking policy, use of entry permits, and appropriate escape and rescue procedures.

(5) Plans to conduct an emergency drill prior to confined-space work to ensure the adequacy of the procedures. A rescue test should be performed to ensure that rescue equipment will fit through the confined-space entrance and to test and practice other confined-space procedures such as communication.

(6) Plans for a stand-by person to be present outside the confined space while workers are inside. The attendant should be trained in the duties of a stand-by person, including appropriate rescue procedures. The stand-by person should have no other duties except to attend the entrance of the confined space, to be in constant communication with the confined-space workers, and to perform a rescue, if needed, with a self-contained breathing apparatus (minimum air supply of 30 min).

(7) Plans to inspect personal protective equipment prior to entry.

(8) Plans for ventilation of the confined space.

(9) Procedures for real-time monitoring of the concentrations of combustible gases or solvent vapors during occupancy.

c. Medical surveillance documentation. Employees required to work with or around solvents, blasting, flame- or arc-spray operations, respiratory equipment, or those exposed to noise above 85 dBA continuous or 140 dBA impact, or those who are required to use respiratory protective devices should be evaluated medically. The contractor should provide a written record of the physical examinations of all employees who may be required to wear a respirator, those who may be exposed to high noise, or those who may be exposed to toxic contaminants. The documentation should include a statement signed by the examining physician confirming that the employees' exams included the following as a minimum:

(1) Audiometric testing and evaluation.

(2) Medical history with emphasis on the liver, kidney, and pulmonary system.

(3) Testing for an unusual sensitivity to chemicals.

(4) Alcohol and drug use history.

(5) General physical exam with emphasis on liver, kidney, and pulmonary system.

(6) Determination of the employee's physical and psychological ability to wear protective equipment, including respirators, and to perform job-related tasks.

(7) Determination of baseline values of biological indices to include:

(a) Liver function tests such as SGOT, SGPT, GCPT, alkaline phosphatase, and bilirubin.

(b) Complete urinalysis.

(c) EKG.

(d) Blood urea nitrogen (BUN).

(e) Serum creatinine.

(f) Pulmonary function tests, FVC, and FEV.

(g) Chest X-ray (if medically indicated).

(h) Blood lead (for those individuals who may be exposed to lead).

(i) Any other criteria deemed necessary by the contractor physician and approved by the Contracting Officer.

d. Respiratory protection program. The contractor should establish and implement a written respiratory protection program that includes instruction and training about respiratory hazards, hazard assessment, selection of proper respiratory equipment, instruction and training in proper use of equipment, inspection and maintenance of equipment, and medical surveillance. The written respiratory program should take into account current and anticipated work conditions for each work area and should be specific for each work area.

e. Air sampling plan. The contractor should prepare and submit plans for conducting air sampling by qualified individuals for toxic contaminants regulated by the Occupational Safety and Health Act (OSHA).

f. Hazard communication program. The contractor should develop and operate a worker hazard communication program for employees in accordance with CFR 29 Part 1910, Section 1200, and state and local worker "right-to-know" rules and regulations. There should be a written program that describes how the employer will comply with the standard, how chemicals will be labeled or provided with other forms of warning, how MSDS's will be obtained and made available to employees and OSHA and NIOSH representatives, and how information and training will be provided to employees. The program should include the development of an inventory of toxic chemicals present in the workplace, cross-referenced to the MSDS file. The written program should also describe how any subcontractor employees and the Contracting Officer will be informed of identified hazards. Specific elements of the program should include:

(1) A file of MSDS's for each hazardous chemical on the chemical inventory, kept in a location readily accessible during each work shift to employees when they are in their work area.

(2) Containers of hazardous chemicals in the workplace should have appropriate labels that identify the hazardous material in the product, have appropriate health and safety warnings, and include the name and address of the manufacturer or responsible party.

(3) Training on:

(a) Provisions of the hazard communication standard.

(b) The types of operations in the work areas where hazardous chemicals are present.

(c) The location and availability of the written program and MSDS's.

(d) Detecting the presence or release of toxic chemicals in the workplace.

(e) The visual appearance, odor, or other warning or alarm systems.

(f) The physical and health hazards associated with chemicals in the workplace.

(g) Specific measures to protect from the hazards in the work areas, such as engineering controls, safe work practices, emergency procedures, and protective equipment.

g. Ventilation assessment plan. The contractor should prepare a written plan for ventilation assessments to be performed by a qualified person for all confined-space work, solvent cleaning, abrasive blasting, and thermal spray operations.

h. Qualifications and experience statement. The contractor should submit a written qualification and experience statement for the person(s) responsible for developing the required safety and health submittals and serving as the contractor's onsite safety and health representative.

i. Safety indoctrination plan. The contractor should submit documentation of the safety indoctrination plan as described in EM 385-1-1, Section 1, "Project Management," Subsection 01.B, "Indoctrination and Training." The documentation should include employee training records for the following items:

(1) The contractor's general safety policy and provisions.

(2) Requirements of the employer and contents of EM 385-1-1 on project safety.

(3) Employer's responsibilities for safety.

(4) Employee's responsibilities for safety.

(5) Medical facilities and required treatment for all accidents.

(6) Procedures for reporting or correcting unsafe conditions.

(7) Procedures for cleaning and surface preparation in a safe manner.

(8) Fire fighting and other emergency training.

(9) Job hazard and activity analysis required for the Accident Prevention Plan.

(10) Alcohol/drug abuse policy.

12-6. Material Safety Data Sheets

The contractor should maintain current MSDS's for all hazardous materials used on the job including cleaning solvents, thermal spray wire or powder, sealers, thinners, and paints or any other product required to have an MSDS as specified in CFR 29 Part 1910, Section 1200. The MSDS's should be available to all contract, subcontract, and government employees.

Chapter 13
Environment and Worker Protection Regulations

13-1. Introduction

Controls are necessary for environmental and worker protection during paint removal. This is relevant to thermal spray to the extent that the removal of aged paint systems containing lead or other hazardous pigments is sometimes performed prior to thermal spray application.

13-2. Regulations

Details on the Federal regulations for environment and worker protection, and a discussion on the means for complying with them, are given in Chapter 11 of EM 1110-2-3400.

Appendix A
References

A-1. Required Publications

Section I
Code of Federal Regulations (CFR)

Title 29 "Labor" contains Office of Safety and Health Administration (OSHA) standards. Copies of CFRs may be obtained from the Superintendent of Documents, U.S. Government Printing Office, Washington, DC 20402.

CFR 29 Part 1910, OSHA
Occupational Safety and Health Standards

CFR 48 1-10.002
Federal Acquisition Regulation - Small Purchase and Other Purchase Procedures

Section II
Federal Agencies

MIL-STD-1687A(SH)
Thermal Spray Processes for Naval Ship Machinery Applications

CID A-A-2962
Enamel, Alkyd

CID A-A-3127
Coating System: Aluminum Epoxy Mastic , For Minimally Prepared Atmospheric Steel

CID A-A-3132
Coating System: Epoxy Primer/Polyurethane Topcoat, For Minimally Prepared Atmospheric Steel

CID A-A-3054
Paint: Heat Resisting (204 deg. C)

TT-P-28
Paint, Aluminum, Heat Resistant (1200 $^{\circ}$F)

TT-P-38
Paint, Aluminum (Ready-Mixed)

TM 5-811-7
Electrical Design, Cathodic Protection

ER 1110-2-1200
Plans and Specifications for Civil Works Projects

EM 385-1-1
Safety and Health Requirements

EM 1110-2-2704
Cathodic Protection Systems for Civil Works Projects

EM 1110-2-3400
Painting: New Construction and Maintenance

ETL 1110-3-474
Cathodic Protection

CEGS-09971
Metallizing: Hydraulic Structures

CEGS 09965
Painting: Hydraulic Structures and Appurtenant Works

Section III
Other Sources

ACGIH-02
American Conference of Governmental Industrial Hygienists. 1996. "Chemical Substances and Physical Agents and Biological Exposure Indices," Cincinnati, OH.

ANSI/AWS A5.01-93
American National Standards Institute/American Welding Society 1993. "Filler Metal Procurement Guidelines," New York.

ANSI/AWS C2.18-93
American National Standards Institute/American Welding Society. 1993. "Protection of Steel with Thermal Sprayed Coatings of Aluminum and Zinc and Their Alloys and Composites," New York.

ANSI Z49.1
American National Standards Institute. 1994. "Safety in Welding, Cutting and Allied Processes," New York.

ANSI Z87.1
American National Standards Institute. 1989. "Practice for Occupational and Educational Eye and Face Protection," New York.

ANSI Z89.1
American National Standards Institute. 1986. "Protective Headwear for Industrial Workers," New York.

ASTM C633-79
American Society for Testing and Materials. 1993. "Standard Test Method for Adhesion or Cohesive Strength of Flame-Sprayed Coatings," Philadelphia, PA.

ASTM D235-95
American Society for Testing and Materials. "Standard Specification for Mineral Spirits (Petroleum Spirits)(Hydrocarbon Dry Cleaning Solvent)," Philadelphia, PA.

ASTM D1200-94
American Society for Testing and Materials. "Standard Test Method for Viscosity by Ford Viscosity Cup," Philadelphia, PA.

ASTM D3734-96
American Society for Testing and Materials. "Standard Specification for High-Flash Aromatic Naphthas," Philadelphia, PA.

ASTM D4138-94
American Society for Testing and Materials. "Standard Test Methods for Measurement of Dry Film Thickness of Protective Coating Systems by Destructive Means," Philadelphia, PA.

ASTM D4285-83
American Society for Testing and Materials. 1993. "Standard Test Method for Indicating Oil or Water in Compressed Air," Philadelphia, PA.

ASTM D4417-93
American Society for Testing and Materials. "Standard Test Methods for Field Measurement of Surface Profile of Blast Cleaned Steel," Philadelphia, PA.

ASTM D4541-95e1
American Society for Testing and Materials. "Standard Test Method for Pull-Off Strength of Coatings Using Portable Adhesion Testers," Philadelphia, PA.

ASTM D4940-89
American Society for Testing and Materials. "Standard Test Method for Conductimetric Analysis of Water Soluble Ionic Contamination of Blasting Abrasives," Philadelphia, PA.

ASTM E337-84
American Society for Testing and Materials. 1996e1. "Standard Test Method for Measuring Humidity with a Psychrometer (The Measurement of Wet- and Dry-Bulb Temperatures)," Philadelphia, PA.

CGA G7.1
Compressed Gas Association. 1997. "Commodity Specification for Air," Arlington, VA.

CGA P-1
Compressed Gas Association. 1991; Interim Amendment 1, 1992; Interim Amendment 2, 1993. "Safe Handling of Compressed Gases in Containers," Arlington, VA.

ISO 8502-6
International Organization for Standardization. 1995. "Extraction of Soluble Contaminants for Analysis - The Bresle Method," Geneva, Switzerland.

NACE #1
National Association of Corrosion Engineers. 1983. "White Metal Blast Cleaning," Houston, TX.

NACE #2
National Association of Corrosion Engineers. 1983. "Near-White Metal Blast Cleaning," Houston, TX.

NACE #5
National Association of Corrosion Engineers. 1996. "Surface Preparation and Cleaning of Steel and Other Hard Materials by High- and Ultrahigh-Pressure Water Jetting Prior to Recoating," Houston, TX.

NFPA 70
National Fire Protection Association. 1993. "National Electrical Code," Quincy, MA.

SAE J827
Society of Automotive Engineers. "Cast Steel Shot," Warrendale, PA.

SSPC-AB 2 1997
The Society for Protective Coatings. "Specification for Cleanliness of Recycled Ferrous Metallic Abrasives," Pittsburgh, PA.

SSPC-AB 3 1997
The Society for Protective Coatings. "Newly Manufactured or Re-Manufactured Steel Abrasives," Pittsburgh, PA.

SSPC-SP 1 1982
The Society for Protective Coatings. "Solvent Cleaning." Pittsburgh, PA.

SSPC-SP 5 1985 (edited 1991)
The Society for Protective Coatings. "White Metal Blast Cleaning," Pittsburgh, PA.

SSPC-SP 10 1985 (edited 1991)
The Society for Protective Coatings. "Near-White Blast Cleaning," Pittsburgh, PA.

SSPC-SP 12
The Society for Protective Coatings. "Surface Preparation and Cleaning of Steel and Other Hard Materials by High- and Ultrahigh-Pressure Water Jetting Prior to Recoating," Pittsburgh, PA.

SSPC-TU 4 1998
The Society for Protective Coatings. 1998. "Technology Update: Field Methods for Retrieval and Analysis of Soluble Salts on Substrates," Pittsburgh, PA.

SSPC VIS 1 1989
The Society for Protective Coatings. "Visual Standard for Abrasive Blast Cleaned Steel," Pittsburgh, PA.

SSPC Paint No. 20 1991
The Society for Protective Coatings. "Zinc-Rich Primers (Type I - Inorganic and Type II - Organic)," Pittsburgh, PA.

SSPC Paint No. 27 1991
The Society for Protective Coatings. "Basic Zinc Chromate-Vinyl Butyral Wash Primer," Pittsburgh, PA.

SSPC Painting Manual
The Society for Protective Coatings. 1995. "Volume 2, Systems and Specifications," Pittsburgh, PA.

SSPC Publication 91-07
The Society for Protective Coatings. 1991. "Effect of Surface Contaminants on Coating Life," Boocock, S.K., Weaver, R.E.F., Appleman, B.R., and Soltz, G.C. Pittsburgh, PA.

A-2. Related Publications

CFR 40, EPA
Code of Federal Regulations. "Protection of the Environment." U.S. Environmental Protection Agency standards.

MIL-D-23003A(SH)
Deck Covering Compound, Nonslip, Rollable.

FHWA 1997
Federal Highway Administration. 1997. "Environmentally Acceptable Materials for the Corrosion Protection of Steel Bridges," FHWA-RD-96-058.

ANSI/NSF 61-1997b
American National Standards Institute/National Sanitation Foundation. "Drinking Water System Components - Health Effects," New York.

SSPC 97-07
The Society for Protective Coatings. 1997. "The Inspection of Coatings and Linings: A Handbook of Basic Practice for Inspectors, Owners and Specifiers," Pittsburgh, PA.

SSPC-SP COM 1995
The Society for Protective Coatings. 1995. "Surface Preparation Commentary," Pittsburgh, PA.

Clemco Industries
Clemco Industries Corporation. 1994. "Blast Off: Your Guide to Safe and Efficient Abrasive Blasting," Washington, MO.

USACERL 1997
U.S. Army Construction Engineering Research Laboratories. "Cavitation- and Erosion-Resistant Thermal Spray Coatings," Technical Report 97/118.

Appendix B
Glossary

A	amperes
Al	aluminum
ACGIH	American Conference of Governmental Industrial Hygienists
AEAC	average equivalent annual cost
ANSI	American National Standards Institute
ASTM	American Society for Testing and Materials
AWS	American Welding Society
cm	centimeter
Cu	copper
CID	Commercial Item Description
CFR	Code of Federal Regulations
CGA	Compressed Gas Association
CEGS	Corps of Engineers Guide Specifications
dBA	decibels
DH	dehumidification
DC	direct current
EM	Engineering Manual
EPA	Environmental Protection Agency
Fe	iron
ft	foot
ft^2	square feet
FDA	Food and Drug Administration
FHWA	Federal Highway Administration

FIFRA	Federal Insecticide, Fungicide, and Rodenticide Act
gpm	gallons per minute
GFCI	ground fault circuit interrupter
HVOF	high-velocity oxygen flame
i.d.	inside diameter
in.	inch
in^2	square inches
ISO	International Organization for Standardization
JRS	job reference standard
kg	kilogram
kPa	kiloPascal
kg/hr	kilograms per hour
$kg/m^2/\mu m$	kilograms per square meter per micrometer
lb	pounds
LD	lock and dam
m	meter
m^2	square meters
$m^2/hr/\mu m$	square meters per hour per micrometer
ml	milliliter
mm	millimeter
mph	miles per hour
μg	microgram
$\mu g/cm^2$	micrograms per square centimeter
μm	micrometer
MSDS	material safety data sheet

MSHA	Mine Safety and Health Administration
NACE	National Association of Corrosion Engineers
NFPA	National Fire Protection Association
NIOSH	National Institute for Occupational Safety and Health
NSF	National Sanitation Foundation
OSHA	Occupational Safety and Health Administration
ppm	parts per million
psi	pounds per square inch
SAE	Society of Automotive Engineers
SSPC	The Society for Protective Coatings
USACERL	U.S. Army Construction Engineering Research Laboratories
USACE	U.S. Army Corps of Engineers
UV	ultraviolet
VOC	volatile organic compound

Appendix C
Summary Description of CEGS-09971, "Metallizing: Hydraulic Structures"

C-1. Required Publications

Appendix C provides a summary description of CEGS-09971, "Metallizing: Hydraulic Structures."

C-2. Part 1, General

Part 1 contains a listing of all applicable references, provisions for payment, definitions, a list of required submittals, safety provisions, and requirements for delivery, storage, and handling of materials and supplies.

C-3. Part 2, Products

Part 2 provides requirements for sampling and testing thermal spray materials and the applied coating. Important requirements include the chemical composition, finish, coil weight, and preparation of metallizing wire. A job reference standard is described with appearance and adhesion requirements.

C-4. Part 3, Execution

Part 3 is the heart of the guide specification. It contains the requirements for surface preparation, metallizing application, workmanship, atmospheric and surface conditions, sequence of operations, approved methods of metallizing, coverage and metallizing thickness, progress of metallizing work, sealing and painting instructions, and the metallizing schedule.

coating. For the most part, the coating delamination had not exposed or resulted in corrosion of the substrate. Film thicknesses measured in the delaminated areas were in the 0.076- to 0.127-mm (0.003- to 0.005-in.) range and are adequate to provide continued protection. Some damage to the zinc-aluminum coating was observed after 8 years, and minor areas of corrosion had begun to appear. It appeared that after 8 years the coating system was approaching the end of its useful life and that replacement would probably be required in 2 to 4 years.

D-2. Racine Locks and Dam

a. Thermal spray system 6-Z-A was applied by wire flame spray to tainter gates at Racine LD during 1994. Recycled steel grit abrasive was used to prepare the substrate prior to application of the thermal spray coating. The blast media was sampled and observed under magnification (30x). It was noted that the media was a mixture of irregular and angular shaped particles. Reportedly, the specified blast profile depth was achieved on the job; however, the profile was probably somewhat rounded because of the blast media mix that was used.

b. Laboratory adhesion tests of the 85-15 zinc-aluminum alloy coating over substrates prepared using different blast media have shown a significant dependence on the shape of the profile and a less significant, but still measurable, dependence on profile depth. Coatings applied over rounded profiles of 0.025 and 0.076 mm (0.001 and 0.003 in.) had average adhesions of 690 and 1980 kPa (101 and 291 psi), respectively. Coatings applied over angular profiles of 0.025 and 0.076 mm (0.001 and 0.003 in.) had adhesions of 7010 and 8230 kPa (1031 and 1210 psi), respectively. Areas of delaminated coating were noted at Racine LD after only 1 year of exposure. The suspected cause of these failures is inadequate surface preparation resulting from the use of a blast media containing significant irregular shaped particles. The irregular shaped particles are too rounded to produce the angular profile needed for good thermal spray adhesion.

c. Cohesive failures and areas of deficient metallizing thickness were also noted. Cohesive failures are most likely due to poor application technique, such as too great of a gun-to-surface distance. Cohesive failures may also have been caused by intercoat contamination or improper equipment setup. These failure causes can be addressed through more thorough inspection and scrutiny of the work in progress. This manual provides additional recommended procedures and methods to address these failures including bend and pull-off adhesion tests and stricter controls on the blast cleaning process.

D-3. Robert C. Byrd Locks and Dam

Thermal spray system 6-Z-A is currently being applied by wire flame spray to new roller gates at Robert C. Byrd LD. Work is scheduled for completion during FY98.

D-4. Pike Island Locks and Dam

a. Portions of the exterior of gate No. 6 at Pike Island LD were metallized with system 6-Z-A in 1995. The metallizing was sealed with vinyl system No. 4. The remainder of the gate was painted with vinyl zinc-rich system 5-E-Z.

b. The gate was inspected during 1996 and several failures were noted. Two large areas (0.929 m^2 and 0.278 m^2 (10 ft^2 and 3 ft^2)) on the downstream skinplate were bare and corroded. In some cases, the thermal spray coating along the edges of the bare areas could be lifted and peeled back for 6.35 to 12.7 mm (0.25 to 0.5 in.). Other small areas of thermal spray were delaminated or damaged. Many of the stiffeners near the bottom of the gate were bare. The bottom 12.7 mm (0.5 in.) of the downstream skinplate was bare metal. Several areas of lifted coating were noted on the downstream skinplate. Presumably, these will eventually be removed by impacting debris. Several scratches in the coating on the upstream skinplate were rusting.

c. The delamination failures are probably due to poor surface preparation, including inadequate surface profile depth and angularity. The skinplate and stiffeners are deeply pitted from previous corrosion. Deep pits are difficult to clean and may have sharp edges that require grinding. Some of the failures are probably caused by the severity of the exposure site, most notably the corroding scratches on the upstream skinplate. The bare metal on the bottom 12.7 mm (0.5 in.) of the downstream skinplate is probably the result of galvanic corrosion between the thermal spray coating and the stainless steel contact imbedded in the concrete sill. This problem is inherent in the design of this and many other dams and is also a problem when standard paint systems are used. Delamination failures on the gate stiffeners could probably be improved by grinding the sharp edges to provide a minimum radius. The stiffeners are also a problem area for paint coatings because the force of impacting debris is concentrated on a much smaller area.

D-5. Olmsted Prototype Investigation

a. A series of prototype dam gates were constructed and installed adjacent to Smithland LD. The prototype gates are a part of a study conducted by the Louisville District in preparation for the design and construction of the Olmsted LD project. The U.S. Army Construction Engineering Research Laboratories (USACERL) performed an evaluation of test coatings at the prototype facility. Coatings were evaluated for corrosion protection and zebra mussel resistance. One of the test coatings was thermal spray system 6-Z-A. The evaluation was conducted during 1996 after 6 months of immersion at Smithland LD.

b. USACERL found that the average film thickness of the metallized coating was 0.635 mm (0.025 in.), exceeding the required average of 0.406 mm (0.016 in.). This gate was not used as a working prototype, but was immersed in the river with the skinplate-side facing down. The skinplate had less than 0.01 percent corrosion, and the downstream side had no visible corrosion . No corrosion was observed on the edges of the gate. The excellent corrosion protection was particularly impressive in light of the mechanical damage evidenced by the deep gouges suffered by this gate during placement and retrieval. The downstream side of the gate has three magnesium anodes. The metallized coating showed a significant amount of blistering, especially along the edges of the gate and in two of the recessed bays on the downstream side of the gate. The worst blistering was found adjacent to two of the three magnesium anodes. Blisters were typically 25.4 to 76.2 mm (1 to 3 in.) in diameter with some being larger. Some of the blisters were broken, although most were intact. There was no observed corrosion associated with the blistering. Blistering may have been caused by the magnesium anodes. The use of sacrificial anodes with significantly higher potential than that of the 85-15 zinc-aluminum alloy coating can produce a current demand close to that of bare steel. This in turn may result in hydrogen evolution and high pH, which may cause blistering of the thermal spray coating. Blistering probably did not occur adjacent to the third anode because of the mud accumulation in this bay. Adhesion of the 85-15 metallizing was excellent except for the blistered areas. The adult zebra mussel population on the metallized portions of the downstream side of the gate (face up side) was only about 1 per square meter. The stainless steel piston rod coupling fixture had in excess of 1000 juvenile zebra mussels per square meter. The average length of the juvenile mussels was about 0.2 mm. The adult zebra mussel density in a small isolated area of the skinplate (face down side) was about 100 per square meter. Based on the size of the adult mussels (10 to 15 mm), it seems likely that they are too old to have settled out on the gate as veligers and instead probably relocated onto the gate from another surface.

c. It was concluded from this study that sacrificial magnesium anodes should not be used in conjunction with thermal spray coatings. It was also observed that 85-15 zinc-aluminum coating provided significant resistance to the settlement of zebra mussel veligers.

D-6. Locks and Dam 53 Investigation

a. Coated test coupons and material samples were placed in test at LD 53 in 1988. Test materials were installed at this sight to investigate the hypothesis that the rate of corrosion is accelerated by some external influence. The results of this evaluation are of interest to the Louisville District because of the replacement dam under construction at Olmsted, IL.

b. Test coatings were evaluated for surface rust, rust undercutting, qualitative adhesion, and blistering. Qualitative adhesion was performed by cutting parallel scribes through the coating using a knife. The coating was lifted from a third cut that was perpendicular to the first cuts. Excellent adhesion equates to no lifting while poor adhesion is complete lifting with no coating elongation. Intermediate values may also be described. Blistering and percent surface rust are described in ASTM standards. The condition of test materials other than coatings is described in general terms. The following compilation shows the condition of the test materials in 1994.

Material or Coating	Condition
Fiberglass grating	no swelling or cracking
Zinc metallizing/epoxy	no defects, adhesion excellent
Galvanized A36 steel	approximately 10-15% rust
A588 steel	general corrosion, 3.17- to 6.35-mm (0.12- to 0.25-in.) thick scale approximates appearance of LD53 pilings, red surface and black underneath
6061 P6 aluminum	10-15% of surface is pitted, black and white corrosion products, jagged pits
E-303/MIL-P-24441	no blistering, adhesion excellent, 3.17-mm (0.12-in.) undercut
MIL-P-24441	surface rust and blisters visible at bolt hole, 3.17- to 4.76-mm (0.12- to 0.18-in.) undercut, adhesion very good
V-766	adhesion excellent, light black sub-film corrosion, blistering at bolt holes, 3.17- to 4.76-mm (0.12- to 0.18-in.) undercut
VZ-108/V-766/V-102	no defects, adhesion excellent
V-106	blistering at bolt holes, adhesion excellent, 1.58-mm (0.06-in.) undercut, dark black sub-film corrosion
V-766/V-103	no blistering, adhesion excellent, light black sub-film corrosion
C-200A	no blistering, adhesion excellent, 3.17- to 4.76-mm (0.12- to 018-in.) undercut
E-303/C-200A	no blistering, adhesion excellent, no undercut
V-766/V-102	no blistering, adhesion excellent, 3.17- to 4.76-mm (0.12- to 0.18-in.) undercut

c. All of the coating systems performed within the expected range. Zinc metallizing topcoated with MIL-P-24441 epoxy, E-303 zinc-rich epoxy primer topcoated with MIL-P-24441, and VZ-108 vinyl zinc-rich primer topcoated with V-766 gray vinyl and V-102 aluminum vinyl were in very good to excellent condition after 6-years exposure at LD53. Any of these coating systems should provide excellent long-term corrosion protection at dam sites on the lower Ohio River. The rest of the coating systems would provide a lesser degree of protection but may still be suitable for some applications.

Subject Index

3

SECTION III

SELECTED SECTIONS FROM

PROCEDURE HANDBOOK FOR SHIPBOARD THERMAL SPRAYED COATING APPLICATIONS

by

National Steel and Shipbuilding Company
San Diego, California

SELECTED SECTIONS FROM

PROCEDURE HANDBOOK FOR SHIPBOARD THERMAL SPRAYED COATING APPLICATIONS

Table of Contents

1. INTRODUCTION

Thermal spray application of aluminum or zinc coatings to a steel substrate is well established for long term corrosion protection. This superior protection has been documented for a wide range of commercial and military applications and in various environments, particularly in the severe marine environment. The U.S. Navy has been specifying the use of thermal sprayed (also known as metal sprayed or flame sprayed) coatings for new construction as well as overhaul and repair work since the late 1970s. To support these coating specifications, the Navy developed and published DoD-STD-2138 (SH) - "Metal Sprayed Coating Systems for Corrosion Protection Aboard Naval Ships" - in 1981. All Navy contractors me required to comply with this document, which outlines stringent procedures for surface preparation and coating application, certification of facilities and operators, production, quality assurance, testing and record keeping. Therefore, any shipyard or contractor that is faced with providing thermal spray services to the Navy should be knowledgeable of all aspects of this DoD Standard.

Thermal spray, although a coating process, is significantly different from conventional painting. Due to the specialized equipment requirements and the relatively low production rates inherent in the thermal spray process, application costs per item or area are usually high, particularly compared to painting. Life-cycle costs however, can be significantly lower for thermal sprayed coatings when compared to painting (Ref. 1). Thermal spray application costs can vary significantly depending on the location or environment of the application (shop vs. shipboard).

Shipyards with Navy contracts requiring thermal spray must decide whether to set up an in-house facility and training program, or subcontract the work to a qualified subcontractor. Most shipyards choose the former alternative to better control cost and schedules, and avoid problems associated with integrating a subcontractor into the busy on-board outfitting construction phase. However, there may be situations where subcontracting is a more cost effective alternative.

Due to unfamiliarity with the processes involved, shipyards new to thermal spray often lack the detailed cost and background information required to perform an adequate make-or-buy (subcontract) analysis. Conversely, shipyards with established thermal spray programs may not possess an adequate documentation trail to enable the periodic review and evaluation necessary to fine-tune their methods and operations.

2. OBJECTIVES

The principal objective of this project is to produce a handbook for thermal spray coating applications. The purpose of the handbook is to provide guidance and assistance to shipyards from two points of view. First, the handbook is intended to guide a shipyard that is preparing to establish a thermal spray program for the first time in accordance with current U.S. Navy requirements. The second objective is to assist shipyards that are currently involved in an active thermal spray program by providing information and data that can be used to analyze and reassess their current methods, thus leading to potential improvements or cost savings.

Since most major shipyards today have established thermal spray facilities to support U.S. Navy contracts, the latter objective will likely be more relevant to the users of this handbook. Significant attention will therefore be placed on making this document useful to those shipyards. In addition, it is the authors' desire that the handbook become a "hands-on," practical tool to benefit the wide range of personnel involved in the thermal spray arena, particularly planners, estimators, engineers, production supervisors and shop managers. The handbook's format and style have been designed with this in mind.

3. PROJECT OVERVIEW

To support the project objectives as discussed in Section 2, this handbook outlines the theory and application of thermal sprayed aluminum (TSA) coatings as well as describing facility requirements, training programs, equipment and application costs, quality, safety and environmental issues. In addition, Section 10 describes and summarizes the key elements necessary to implement a shipyard thermal spray program.

Although aluminum, zinc and various alloys are used successfully in the thermal spray process, this handbook addresses aluminium coatings only, since this coating is currently being specified by the U.S. Navy for shipboard corrosion control. Also, all information presented herein supports the requirements of the currently approved government standard for thermal spray, DoD-STD 2138. Some information is presented in the form of direct quotes from the standard, and the complete standard is included as Appendix H. It should be noted that at the time of this writing a revision to 2138 is nearly completed and expected to be issued by the Navy early in 1992 (2138-A). This handbook makes several references to updated information contained in the draft version of 2138-A.

The research for this project was accomplished in several phases as Summarized below

1 Conduct shipyard and subcontractor surveys related to methods, production rates and equipment. (Survey results are summarized in Appendix A.)

- Identify and analyze current methods of thermal spray applications and related equipment usage.
- Compile production rate data for various types of thermal spray applications.
- Perform an industrial engineering study to determine overall costs per unit area.
- Outline requirements and procedures leading to certification in accordance with DoD-STD 2138.
- Assemble all data and information to create the thermal spray handbook

Also, visits were made to the following shipyards to tour thermal spray facilities and discuss the project surveys:

- Avondale Shipyards, New Orleans, LA.
- Bath Iron Works, Bath, ME.
- Ingalls Shipbuilding, Pascagoula, MI.
- Puget Sound NSY, Bremmerton, WA.
- SIMA, Naval Station, San Diego, CA.

4. INTRODUCTION TO THE THERMAL SPRAY PROCESS

4.1 Definition of Thermal Spraying

Thermal Spraying is a group of processes in which finely divided metallic or nonmetallic materials are deposited in a molten or semi-molten condition on a prepared substrate to create a spray deposit. The material used may be in the form of powder or wire. The thermal spraying gun generates the necessary heat by using combustible gases or an electric arc. As the materials are heated they change to a plastic or molten state and are accelerated by a compressed gas. The confined stream of particles is conveyed to the substrate. The particles strike the surface, flatten and form thin platelets (splats) that conform and adhere to the irregularities of the prepared surface and to each other. As the sprayed particles impinge upon the substrate, they cool and build up, particle by particle, into a lamellar structure, thus a coating is formed (see Fig. 4.1).

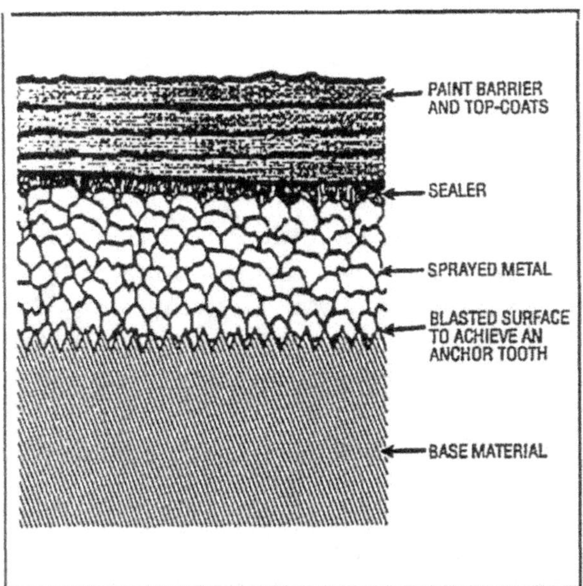

Fig. 4.1 Typical Coating Cross Section

The commercial use of thermal spraying covers many applications such as: restoration of dimension, hard facing, electrical conductivity, electrical resistance, wear resistance, thermal barriers, Radio Frequency Interference (RFI) or Electro-magnetic Interference (EMI) shielding and corrosion protection to a variety of components. It is important to note that there is a wide utilization of thermal spray coatings for applications in nearly all industries. Thermal spraying is a time-proven process that when properly applied and engineered provides a surface treatment which can greatly enhance the serviceable life of components.

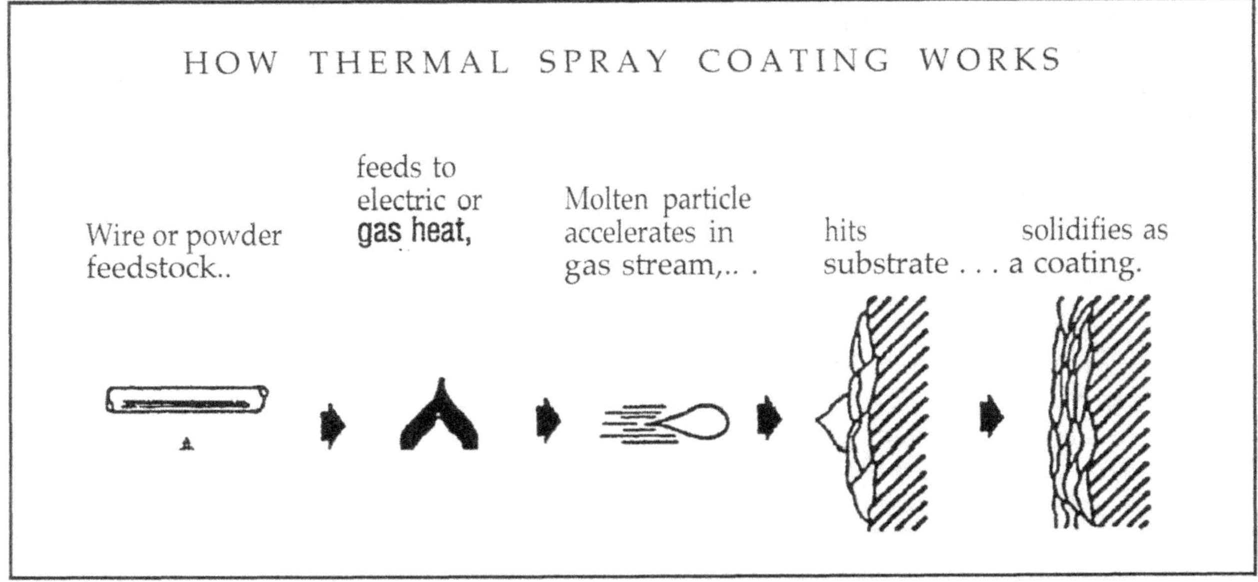

4.2 History of TSA Coatings for Corrosion Protection

Most U.S. shipyards are currently, or soon will be using thermal sprayed aluminum (TSA) coatings to provide corrosion control of various components for the U.S. Navy. Following is an overview of where TSA coatings are currently being applied in industry. This information is important background that shows the acceptance of these coatings throughout all types of industries for corrosion protection.

Thermal sprayed coatings are used for protection of iron and steel in a range of corrosive environments. Long-term effectiveness (twenty years) in both industrial and marine locations has been documented (Ref. 2). The two most commonly used processes to apply these coatings are Combustion Flame Spraying and Electric Arc Spraying. Flame and arc spraying give rapid, uniform coverage at economical spray rates, and are replacing conventional paint primers for many corrosion applications. The natural surface texture of these coatings provides an excellent base for sealers or paint top coats.

Zinc and aluminum coatings provide the broadest atmospheric protection, and the choice in their use is often based on the relative ease of application and comparative costs for the specific case. These coatings are more corrosion resistant than steel, and hence form an effective atmospheric barrier. Zinc and aluminum are anodic to steel. They also protect the ferrous substrate in electrolytic solutions. The coating serves as a sacrificial anode and is consumed over time. This reserve cathodic protection acts to prevent corrosion of the substrate even when the coating coverage is incomplete or suffers mechanical damage.

Aluminum corrodes less rapidly than zinc in highly acidic conditions, while zinc performs better than aluminum in alkaline conditions. Aluminum and zinc coatings both have good adhesion to grit-blasted steel. Because thermal spraying does not cause excessive heating of the substrate, there is no effect on the mechanical properties of the substrate.

There is a history of corrosion protection by thermal spraying for structural steel work. Included are buildings, bridges, towers, radio and TV antenna masts, steel gantry structures, high power search radar aerials, overhead walkways, railroad overhead line support columns, electrification masts, tower cranes, traffic island posts, and street and bridge railings. Wellhead assemblies for offshore use have been coated for salt atmosphere corrosion protection since the 1950s. The process has been used for flare stacks, refinery columns, and for external protection of oil and propane gas storage tanks.

The interiors of fluid cargo rail cars are thermal sprayed to control fluid purity and guard against iron contamination. Steel railroad cars are zinc sprayed for corrosion protection. These coatings should last the lifetime of the cars, thus eliminating the need of removal from service for painting (approximately every five years). Spraying has been used to protect pipelines against many types of environments. Lengths up to forty feet (12m) have been successfully coated internally. Pipe couplings, manhole covers, and other small industrial items are also coated.

In marine applications, hulls, deck sections, and portions of barge, scow, tug, and fishing vessel superstructures have been sprayed with excellent long term results. Lifeboats and floating caissons have also been coated, as well as smaller items ***such*** as ship rudders and the axles of boat trailers. A common usage of metal spray is on piers, pilings, and ferry berths.

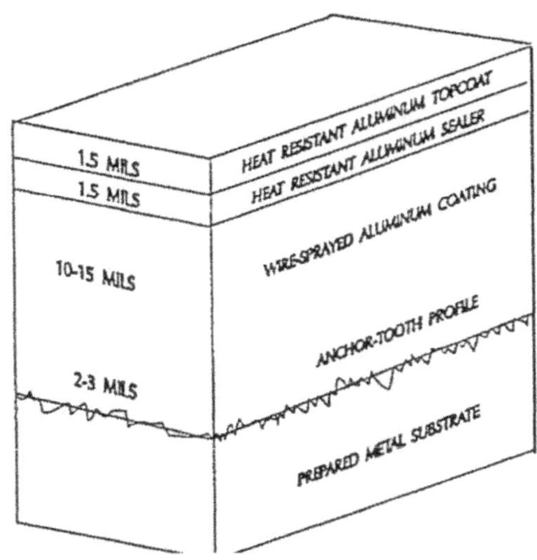

Fig. 4.2 Type I System for High Temperature Applications

One example of process problems is found with surface preparation requirements. The Standard requires a white metal surface with an anchor tooth pattern of 2-3 mils. This is measured with a press-o-film profile tape. The steel surface is required to be completely thermal sprayed within six hours of surface preparation. When the correct anchor tooth pattern is not achieved or the steel surface is allowed to remain uncoated longer than six hours, lower bond strengths are obtained. This can result in coating failure in less than one year.

A thermal sprayed aluminum (TSA) coating applied in accordance with DoD-STD 2138 will have a non-through porosity thickness of about seven roils and a tensile strength greater than 2000 psi. Seven roils will give barrier protection and anodically protect the substrate if the coating is damaged, exposing the base material to the environment. Additionally, a seal coat is always applied to a TSA coating to improve the barrier and provide a tie coat for subsequent paint coats. The Navy's TSA coating system is essentially a metallized primer topped with the designated top coat paint system. The TSA component is the long-lived component, and the five-coat paint component is renewed as required by "wear and tear" of the shipboard traffic and the marine environment. TSA coatings can provide effective long-term (over ten years) corrosion protection to steel in shipboard applications.

The U.S. Navy has two aluminum sprayed coating systems tailored to the operational environment the coated surfaces will see in service. Type I system provides for high temperature applications [surface operating temperatures greater than 79 degrees C. (175 degrees F.)] such as steam valves, and requires an aluminum coating thickness of 10-15 roils (0.010 to 0.015 inches) sealed with a heat resistant paint (Fig. 4.2).

The Type II system is for low temperature systems [surface operating temperature less than 79 degrees C. (175 degrees F.)] and is used for selected marine topside and interior applications. The Type II system consists of 7-10 roils (0.007 to 0.010 inches) of aluminum coating sealed with thinned epoxy polyamide paint (MIL-P-24441 /1) and topcoated with two additional layers of epoxy polyamide paint (Fig. 4.3).

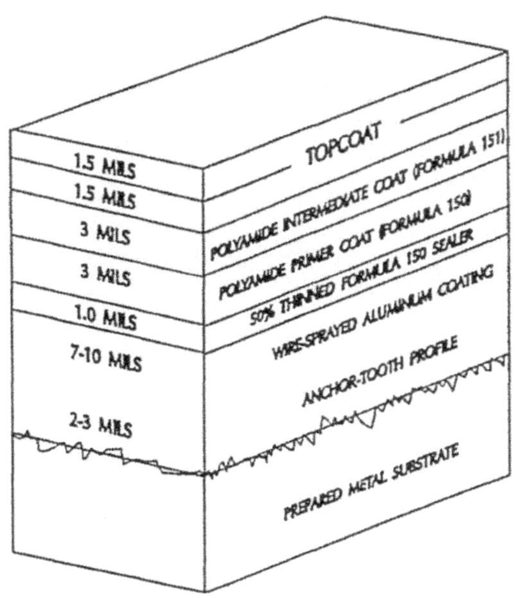

Fig. 4.3 Type II System for Low Temperature Applications

It is important to point out that the sealer/epoxy polyamide paint is chosen to make the aluminum system compatible with U.S. Navy painting practices. Type II systems for Navy ship application as shown in Fig. 4.3 require one additional camouflage/color topcoat. This topcoat can be one coat of MIL-P-24441 epoxy or alkyd haze grey enamel (TT-E-490). TT-E-490 is applied in two thin coats -1.5 mils per coat. Other topcoats may be required by individual ship specifications, or local air quality regulations.

4.4 Approved Applications per DoD-STD-2138

The following is a list of approved applications of metal spray coatings for corrosion control on Navy surface ships

Category I.
Machinery space components
(Type I, High Temp)

(a) Aluminum coating -10 to 15 mils thick
- (1) Low pressure air piping
- (2) Steam valves, piping and traps (except steam turbine control valves)
- (3) Auxillary exhaust (such as stacks, mufflers, and manifold)
- (4) Air ejection valves
- (5) Turnstile

Category II.
Topside weather equipment
(Type II, Low Temp)

(a) Aluminum coating -7 to 10 roils thick
- (1) Aircraft and cargo tie downs
- (2) Aluminum helo decks
- (3) Stanchions
- (4) Scupper brackets
- (5) Deck machinery and foundations
- (6) Chocks, bits, and cleats
- (7) Pipe hangers
- (8) Capstans/gypsy heads (except wear area)
- (9) Rigging fittings (block and hooks)
- (10) Fire station hardware
- (11) Lighting fixtures and brackets

Category III.
Interior wet spaces
(Type II, Low Temp)

(a) Aluminum coating -7 to 10 roils thick:
- (1) Decks in wash rooms and water closets
- (2) Pump room deck and equipment support foundations
- (3) Fan room decks and equipment support foundations
- (4) Water heater room decks and equipment support foundations
- (5) Air conditioning room decks and equipment support foundations
- (6) Deck plate supports
- (7) Machinery foundations
- (8) Boiler air casings (skirts)

4.5 List of Typical Shipboard TSA Coated Items

It is important to the success of the existing Navy corrosion control program and long term utilization that each shipyard use its experience and expertise to continually examine and suggest additional uses and applications for the thermal spray aluminum coatings systems. The following list represents some standard items that have been preserved using thermal sprayed aluminum coatings.

Arm, FAS Swivel
Assembly, FAS Swivel
Armbar, Chair
Base, P-250 Box
Blower, Soot
Bolt, Baxter
Box, P-250
Brace, Accom-Ladder
Bracket/Pipe, Light
Bracket, Bottle-Rack
Bracket, CO_2 Bottle
Bracket, Helo-Net Securing
Bracket, Scupper
Bracket, Searchlight
Controller, Capstan
Counterweight, Director
Coupling Pipe
Cover, IMC Speaker
Cover, Cable BPSMS
Cover, Chain Locker
Cover, Chock
Cover, Edge Light
Cover, Fuel Oil Vent
Cover, Handwheel
Cover, Hawse Pipe
Cover, Junction Box
Cradle, Life Raft
Davit, Portable
Director, Counterweight CWL
Dog, Wt Door
Door, Watertight
Ducting
Eye, Bolt Lifting
Fitting, NATO
Foundations, Machinery
Flange, Blank
Frame Net
Frame, Stanchion
Guard, Winch Safety
Handle, Strainer
Handrail, Debark
Handwheel, Various
Hatch (Large)
Hinge
Housing, Reel Shaft
Ladder, Vertical
Locker, Pyro
Manifold
Mount, FAS Bulkhead
Mount, Reel Line
Mount, Saluting Gun
Pad, Fairlead
Piping, FAS
Plate/Bracket, Light
Plate, Fuel Gage
Rack, Bottle Gas
Reel, Line
Relief Exhaust
Rod, Reach
Roller, Ramp
Roller, Stopper
Screen, Bullnose
Scuttle
Shackle, Unrep
Sheave, Fairlead
Shield, 50 Cal
Sign, Parking
Socket, Portable Davit
Stanchion
Valves, Globe
Valves, Governor
Valves, Reducer
Valves, Regulating
Vent, Fuel Oil
Wrench Anchor
Yoke, 50 Cal
Yoke Searchlight

4.6 Prohibited Applications Per DoD-STD 2138

Metal sprayed coatings for use in U.S. Navy corrosion control applications are intended for selected application to steel and aluminum surfaces. TSA coatings for corrosion control should not be used for the following

l Plastic, rubber, painted surfaces

- Internal surfaces of moving machinery (example pump casings, valves, etc.)

Ž Brass, bronze, copper-nickel or monel surfaces

- Stainless steels, 17-4PH, 15-4PH

- Surfaces subject to strong acids or bases (example aircraft catapult slides)

- Threads of fasteners, and valve stems

- Within ¾″ of surfaces to be welded

- Steel alloys with yield strength greater than 120,000 lb/in^2

- Nonskid deck coatings (except as approved by NAVSEA for research and development evaluation)

l Exterior underwater hull surfaces

Ž Sanitary tanks interior

5. THERMAL SPRAYED COATING APPLICATION

5.1 The Thermal Sprayed Aluminum Process

The Thermal Sprayed Aluminum (TSA) coating system consists of a properly prepared substrate to which a thermal sprayed aluminum coating is applied and subsequently sealed with organic paint, and top coated with paint as required to meet barrier, functional, or cosmetic requirements.

DoD-STD 2138 describes two TSA coatimz systems; one for components that experience high temperature operation conditions and the second for ambient conditions. Refer to Section 4.3 for system descriptions.

TSA coatings may be applied by using the combustion flame or electric-arc thermal spray process as discussed later in this section. The production process for applying the TSA coating system can be divided into five major production steps. Table 5-A summarizes the production work station functions and quality control (QC) checkpoints. Process control is further discussed in Section 7.

WORKSTATION	FUNCT ONS	
	PRODUCTION	QC CHECK
1. Surface Preparation	• Solvent clean • Trisodium phosphate wash • Abrasive cleaning • Heat cleaning • Caustic bath	1 Clean substrate
2. Masking	• Mask surfaces not to be coated	Ž Properly masked
3. Anchor-tooth Blasting	• 2-3 mil anchor-tooth • White-metal finish	• Al_2O_3 grit quality/size • Clean, dry air • SSPC-5 finish
4. Thermal Spraying	• Preheat substrate • Spray 90-45° to substrate and 5-8 in. standoff - 3-4 mil per crossing pass - 7-10 mil for <175°F service; 10-15 mil for >175°F service	• Substrate >10°F above dew point • Proper time between anchor-tooth blasting and start/-completion of spraying • Coupon bend test (equipment system check) • Thickness measurement • Visual examination
5. Sealing/ Painting	• ≤175°F service: 5-coat paint schedule; 10 mil total thickness • >175°F service: 2-coat paint schedule; 3 mil total thickness	• Wet film thickness • Minimum/maximum drying times between paint coats • Dry film thickness

Table 5-A Work Station Function and Quality Control Checkpoints

Items to be TSA coated are transferred by production personnel to various work stations for surface preparation, decreasing and masking, strip anchor-tooth blasting, thermal spraying and painting. Items to be TSA coated are anchor-tooth blasted to white metal with a 2-3 mil anchor-tooth pattern. Following anchor-tooth blasting, items are transferred immediately, generally within a few minutes, to the TSA coating station. The component is then aluminum thermal sprayed, sealed and painted in accordance with DoD-STD 2138 and ship specifications.

The Process Flow Chart in Fig. 5.1 represents a typical production flow to accomplish the TSA coating system. Numbers in parentheses refer to related sections of DoD-STD 2138.

Each facility that intends to apply TSA coatings should develop an internal industrial process instruction that details all the production and quality control checkpoints, such as equipment, consumable materials, safety, quality control, operator training and certification. The Navy's military standard for applying the TSA coatings is the primary technical reference for this process instruction. (See Appendix H).

NASSCO developed such a document during the early stages of their thermal spray program. This "Flame Spray Manual" was prepared by the Quality Assurance Department (Ref. 3). Following is an outline of the contents:

- Certification Requirements
- Equipment/Material Requirements
- Safety
- Surface Preparation
- Process Instructions
- Inspection/Testing Requirements

5.2 Types of Equipment for TSA Coating Applications

The thermal spraying processes and related equipment for the application of aluminum for corrosion control can be divided into two basic categories combustion and electric arc.

Combustion Spraying

Thermal spraying utilizing the heat from a chemical reaction is known as combustion gas or flame spraying (oxygen and a fuel gas). Any substance that does not sublime and that melts at temperatures less than 5000 F. (2760 C) may be flame sprayed. The materials used are metals (and alloys) in the form of wire or powder.

The equipment for thermal spraying with wire is similar to that shown in Fig. 5.2, and a typical wire thermal spraying gun cross section is shown in Fig. 5.3.

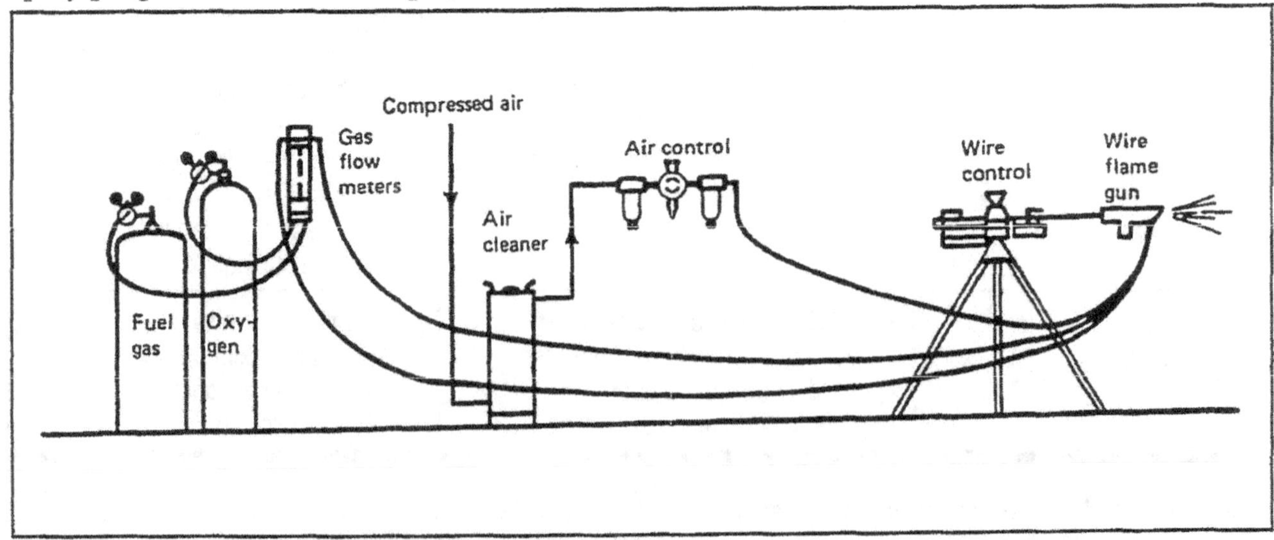

Fig. 5.2 Typical Combustion Wire Installation

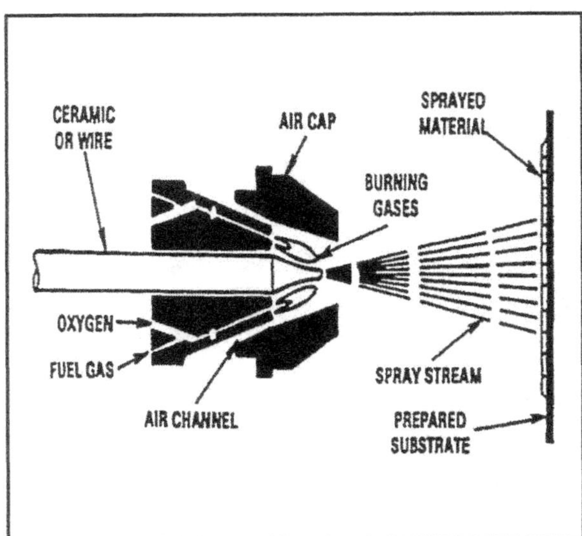

Fig. 5.3 Cross Section of Typical Combustion Wire Gun

The feedstock material (wire) is drawn by drive rolls into the rear of the gun. The rolls are powered by an electric motor, an air motor, or an air turbine. The feedstock proceeds through a nozzle where it is melted by the coaxial flame of burning gas.

One of the following fuel gases may be combined with oxygen for use in flame spraying acetylene, methylacetylenepropadiene stabilized (MPS), propane or natural gas. Acetylene is widely used because higher flame temperatures are attainable. However, in many cases, lower temperature flames can be used with economic advantages. A fuel gas flame is used for melting only, and not for propelling or conveying the coating material. To accomplish spraying, the flame is surrounded with a stream of compressed gas, usually air, used to atomize the molten material and to propel it onto the substrate. In special applications, an inert gas maybe used in lieu of air.

Powder flame spray guns are lighter and more compact than other types of thermal spraying equipment. The powder feedstock may be a pure metal, an alloy, a composite, or any combination of these. The feedstock is stored in a hopper that may be integrated with, or connected to the gun. A small amount of gas is diverted to carry the powder into the oxygen-fuel gas stream, where the powder is melted and carried by the flame onto the substrate. The general arrangement of an installation for powder flame spraying is shown in Fig. 5.4, and a typical gun cross section is shown in Fig. 5.5. Variations in the powder flame spraying process include compressed gas to

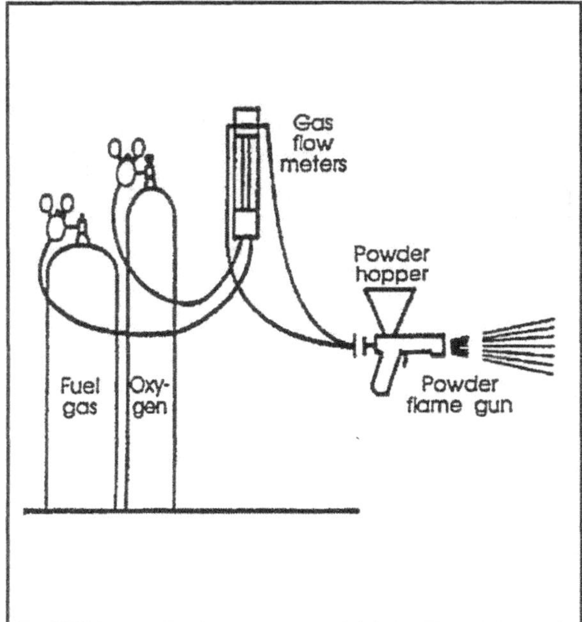

Fig. 5.4 *Typical Combustion Powder Installation*

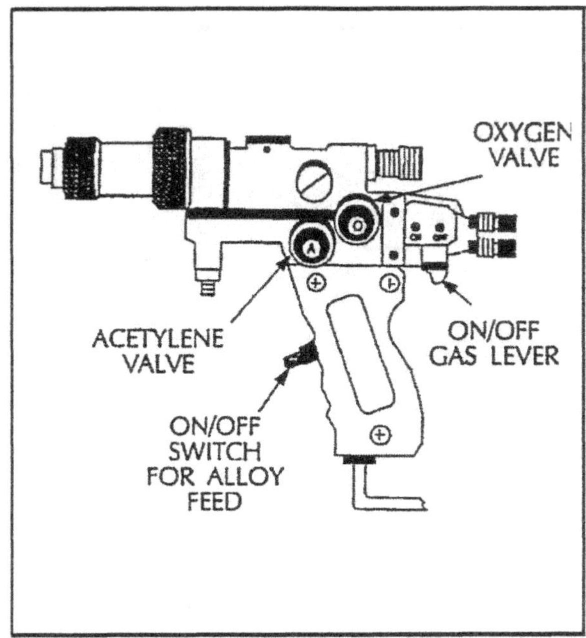

Fig. 5.5 Typical Combustion Powder Gun

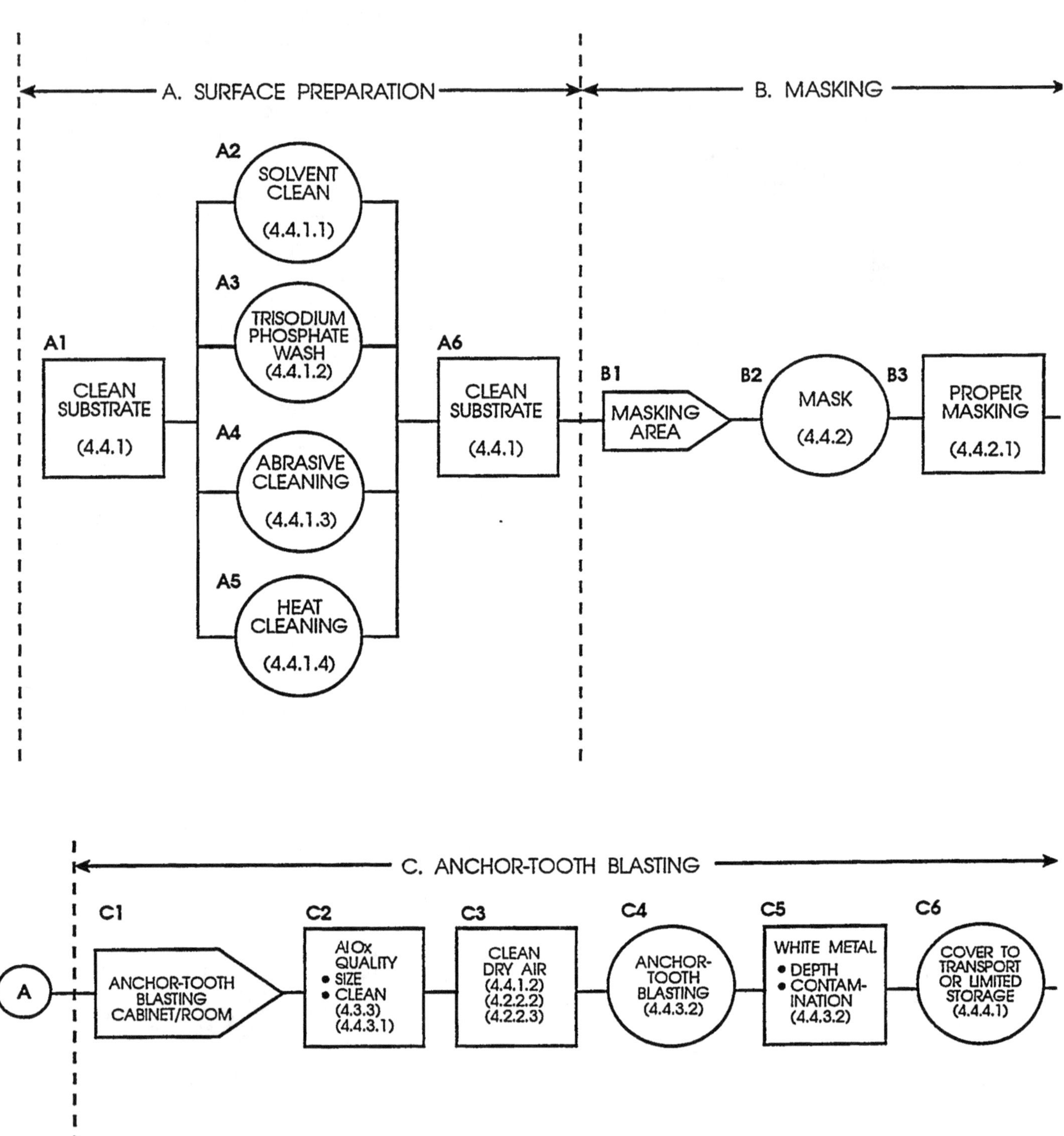

Fig. 5.1 Process Flow Chart

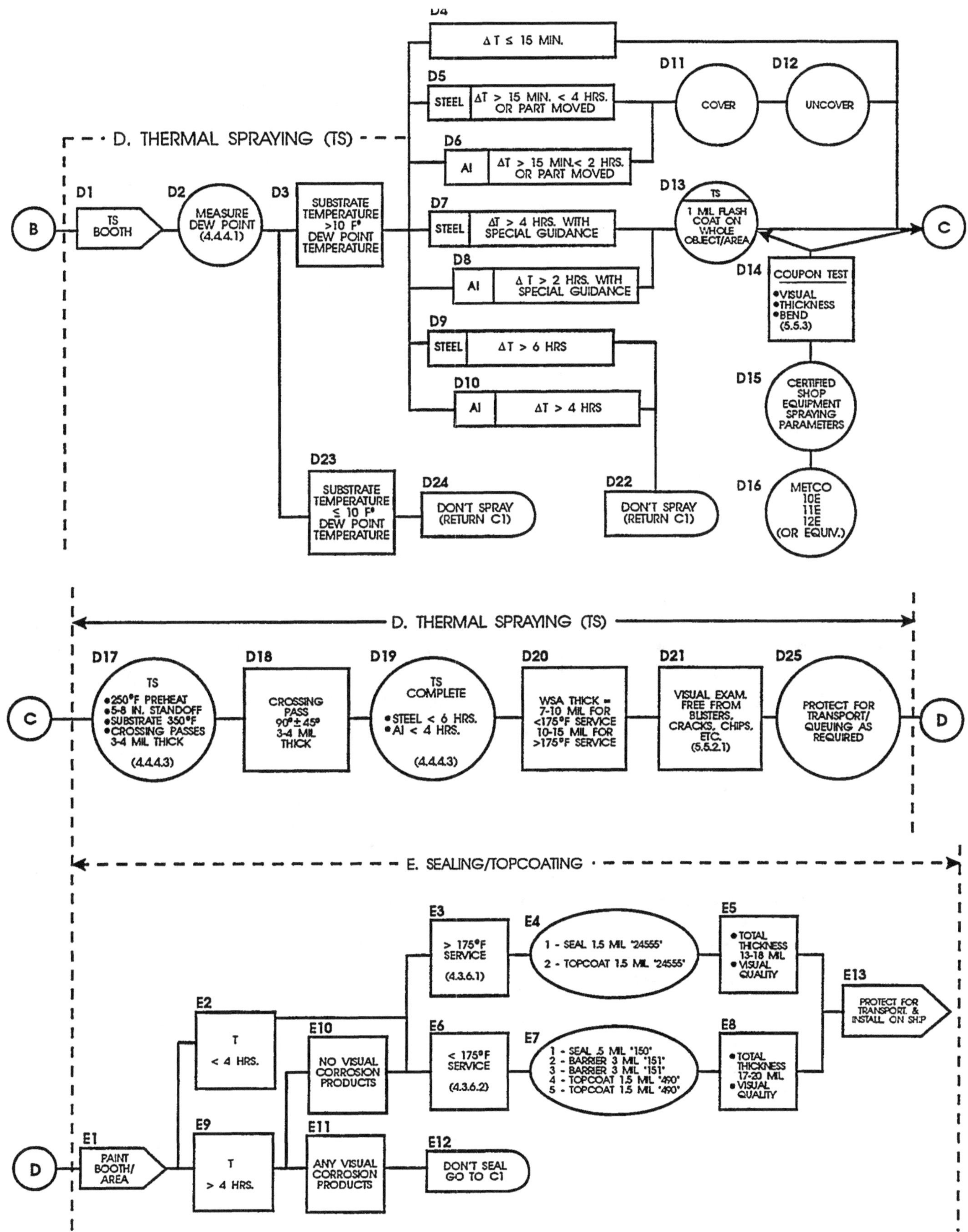

Fig. 5.1 Process Flow Chart (continued)

feed powder into the flame, additional air jets to accelerate the molten particles, a remote powder feeder with an inert gas to convey powder through a pressurized tube into the gun, and devices for high speed acceleration at atmospheric pressure. Such refinements tend to improve powder flow rate, and sometimes to increase particle velocity, which enhances bond strength and spray deposit density.

In all thermal spraying processes, the particle velocity affects the structure and the deposit efficiency of the coating. If the particle velocity is too low, some particles may be volatized and result in coating deterioration and elevated operating costs. If the wire or powder are not properly heated, deposit efficiency will decrease rapidly, and the coating will contain trapped, unmelted particles. Table 5-B compares the equipment working temperatures and particle velocities for each System.

THERMAL SPRAY EQUIPMENT TYPE	PARTICLE VELOCITY, FPS	FLAME WORKING TEMPERATURE, F
Gas Combustion Heating		
Powder flame Spray	80-120	4600-4800
Wire Flame Spray	800	5000-5200
Electric Arc Heating		
Electric Arc spray	800	10,000-12,000

Table 5-B Particle Velocity Comparison

Electric Arc Spraying

In the wire arc process, two consumable wire electrodes, that are at first insulated from each other, automatically advance to meet at a point in an atomizing gas stream. A potential difference of 18 to 40 volts applied across the wires, starts an arc that melts the tips of the wire electrodes. An atomizing gas, usually compressed air, is directed across the arc zone, shearing molten droplets which form the atomized spray. The arc spray system is comprised of components as illustrated in Fig. 5.6.

The arc spray gun is illustrated in Fig. 5.7. Wire electrodes are fed through wire guides and into the contact tips. The atomizing nozzle conducts the compressed air and directs it across the arc zone. Insulated power cables connect the gun to the power source. Arc guns also include mechanisms for feeding the wire at a controlled rate. Contact tips are sized for a particular wire diameter. ON and OFF switches are provided on the gun to control the wire feed, compressed air supply, and electric power supply. The arc temperatures considerably exceed the melting point of the spray material. During the melting cycle, the fed wire is super heated to the point where some volatilization may occur. The high particle temperatures produce metallurgical interactions or diffusion zones, or both, after impact with the substrate.

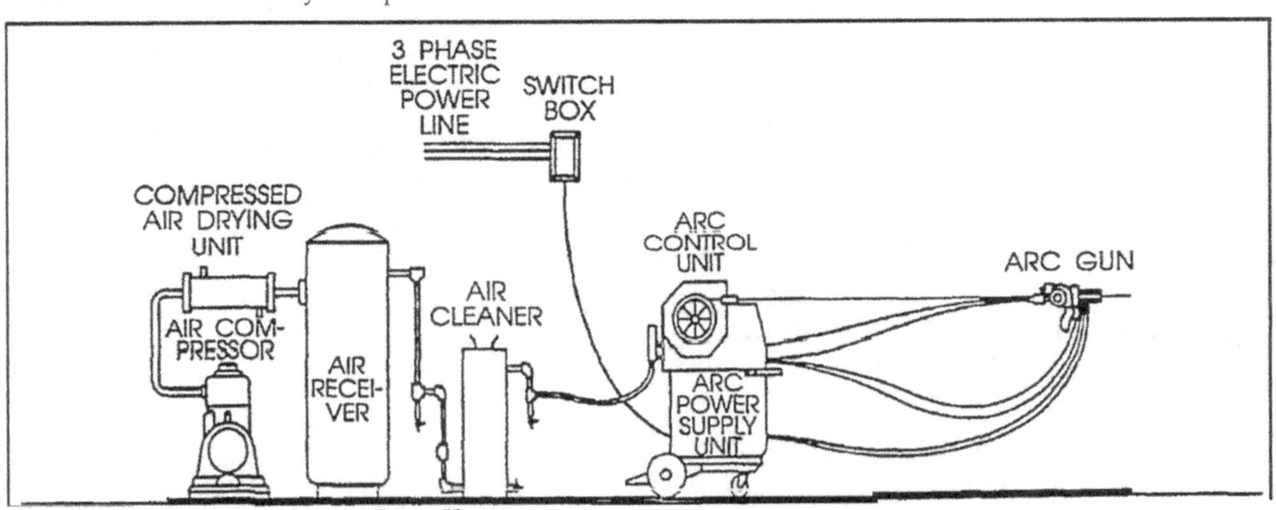

'ig. 5.6 Typical Arc Spray Installation

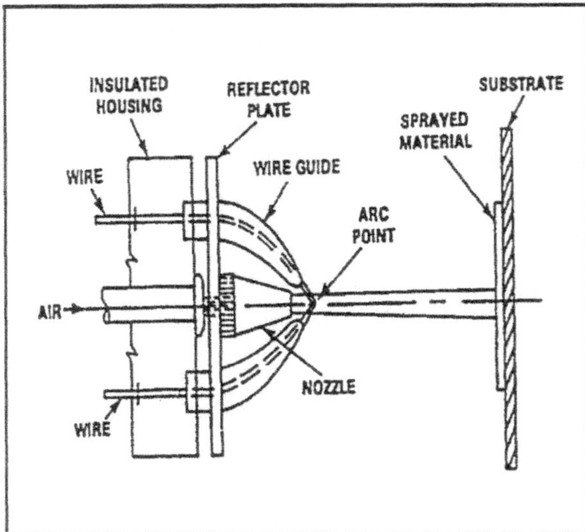

Fig. 5.7 Cross Section Schematic of an Arc Spray Gun

These localized reactions form minute weld spots with good cohesive and adhesive strengths.

The arc process normally has higher spray rates than other spray processes. Factors controlling the application rate are the current rating of the power source and the permissible wire feed rate to use the available power. Section 9 discusses the economics of the arc and combustion processes.

Typical arc wire systems require DC power source providing a voltage between 18 to 40 volts. Constant potential power sources are usually used. The arc gap and spray particle size increase with a rise in voltage. The voltage should be kept at the lowest level, consistent with arc stability, to provide smooth, dense coatings and improved deposit efficiency.

The arc wire control unit is comprised of two reel or coil holders, which are insulated from each other. Wires of larger diameters are usually in coil form, while smaller diameter wires are preferably layer level wound on reels or in barrels. The unit is connected to the gun by flexible insulated cables.

The control console incorporates the switches and regulators necessary for controlling and monitoring the operating circuits that power the gun.

5.3 Surface Preparation Procedures

Surface preparation is the foundation of a correctly applied thermal sprayed coating. Coating adhesion is directly related to the cleanliness and the roughness of the prepared surface. Adherence to qualified procedures in preparing a surface is necessary to ensure successful application of the thermal spray coating.

Pre-Cleaning

The first step in the preparation of a substrate for thermal spraying is to remove all surface contaminants such as scale, oil, grease and paint. The heat of spraying will not remove contaminants, and contamination will inhibit bonding. After all contaminants have been removed, the cleanliness should be maintained until the spray cycle has been completed.

Hot vapor degreasing is a common, economical, and efficient method for removing organic contaminants. Parts should be soaked fifteen to thirty minutes to remove oil from surface pores. Porous materials, such as sand castings or cast irons, should be soaked for longer periods. If objects are too large for vapor decreasing, steam cleaning, submerging in hot detergent solutions, or manually cleaning with a solvent that is oil free may be required. Chlorinated solvents leave a slight residue that can be removed by immersion washing or wiping with isopropyl alcohol. Residue should be mechanically removed.

The use of common degreaser solvents such as perchlorethylene, trichlorethylene and 1,1,1 trichlorethane, is not currently recommended due to more stringent Federal and local air quality regulations. Most hydrocarbon solvents are hazardous, and manufacturer's instructions should be followed regarding usage, location and disposition. Investigate the local EPA or Hazardous Waste Regulations for use and disposal of solvents before selecting the materials. Also, the potential recycling of solvents should be investigated with the supplier. Non-hazardous, biodegradable cleaners or detergents should be used whenever possible.

Various components manufactured from porous materials such as castings, may absorb considerable quantities of oil. This oil may bleed out during a subsequent spraying operation, even after precleaning using vapor decreasing or solvent or detergent washing. Welded assemblies that have received fluorescent penetrant inspection are particularly subject to this condition. Oven baking at 600 degrees Fahrenheit for four hours dries the oil and prevents bleed out.

Ultrasonic cleaning can be used when contaminants are lodged in confined areas. The equipment consists of a holding tank for the cleaning solution and a source producing ultrasonic vibrations within it.

Strip Blast Cleaning

Dry abrasive blasting, commonly described as strip blasting, is an effective method for removing baked-on deposits, scale, paint or oxides. Abrasive blasting is accomplished by directing a compressed air steam containing abrasive particles through a nozzle to the surface. The blasting operation should be conducted by equipment other than the unit assigned for the final (anchor-tooth) blasting of substrates for spraying. This prevents contamination of the blasting materials. The object of this cleaning technique is to ensure there are no contaminants remaining on the surface that will prevent maximum adherence of the thermal sprayed aluminum coating.

Masking Techniques and Materials

For some components, only specific areas of a workpiece may need to be coated. (Section 4.6 defines prohibited areas of TSA application) In such cases, there are areas that must be protected, both from unnecessary surface roughening and from adherence of the sprayed material. The technique of protecting the areas not to be blasted or sprayed is called masking. Different masking materials and techniques may be required for grit blasting and for spraying, though commonly the same mask may be used for both operations. Some masking materials used with success are listed in Table 5-C.

For most grit-blast masking applications, ordinary rubber tape or duct tape, applied several layers thick, affords satisfactory protection to surfaces. These materials have the advantage of being inexpensive and easy to cut or shape as needed. Adhesive-backed silicon tapes provide the required protection for abrasive blasting and thermal spraying, and are less likely to be damaged by abrasive blasting and spraying than duct or rubber tapes. However, silicone tapes are approximately five times more expensive than duct tape.

Metal or rubber shields may be used, provided the geometry of the workpiece is not too complex. Of the two materials, rubber is the better choice as a grit blasting mask because it is not cut by the grit blast abrasive, whereas a metal mask will ultimately erode. However a metal grit blasting mask may be used as a thermal spray mask as well, while rubber, except silicon, cannot. Provisions must be made, when using metal masks, to prevent coating build-up on them. This may require that the masks are coated with a suitable "stop-off' material, or equipped with some mechanical means of removing deposited material. The use of metal or rubber shields is generally limited to flat or cylindrical surfaces.

DESCRIPTION	PRODUCT IDENTIFICATION	FORM	METHOD OF APPLICATION	FOR GRIT BLASTING	FOR COATING	METHOD OF REMOVAL
Silicon resin backed tape	Bron 18647	Tape	Manual	X	X	Peel
Fiber glass tape	Bron BT1077 Insulectro 06004	Tape	Manual		x	Peel
Vinyl tape	3M Scotch #33	Tape	Manual	x		Peel
Liquid masking	Metco Antibond	Liquid	Brush		x	Water wash or brush
Commercial grade Silicon	Silicone	Tubing, sheeting, plugs	Cut as templates and place manually	x	x	–
Liquid masking	Tafa Spray Guard	Liquid	Brush Dip	x	x	Peel
Duct tape	—	Tape	Manual	x	x	Peel
Metallic masks	—	Steel-12-16 ga.	cut to fit	x	x	

Table 5-C Description Of Masking Materials

Another masking technique suitable for thermal spray applications is the use of liquid masking compound. These compounds, when applied to areas adjacent to the desired deposit area, act as stop-offs to prevent the adherence of the sprayed materials to the base metal. They may be applied by painting or dipping prior to or following grit blasting, making them easy to use. Threaded holes, slots, and key ways can be protected from thermal spraying by metal, silicon rubber, or carbon inserts.

Anchor-Tooth Blasting

After cleaning and masking, abrasive blasting is used to produce a surface that will maximize the adherence of the sprayed aluminum coating. Proper roughening is as important as cleaning. During thermal spraying, the plasticized molten particles form as platelets upon impact with the substrate. The platelets, as they cool and harden, must adhere to the prepared surface that will provide the mechanical adherence (bond strength).

Surface roughening by abrasive blasting is used to strengthen both the coating and the bond by

- providing compressive surface stresses
- interlocking laminations (or layers)
- increasing the bond area
- decontaminating the surface

The degree of roughness and general surface condition required to produce a receptive surface for the thermal sprayed aluminum coating is defined in DoD-STD2138 as follows

• The blasted surface will have a white metal blast appearance with an anchor-tooth (not peened) surface profile of two to three roils and validated with profile tape and a dial micrometer.

• Blasting will be done according to the requirements of SSPC No. 5 (or NACE No. 1). A white metal blast cleaned surface finish is defined as a surface with a gray-white, uniform metallic color, roughened to form a suitable anchor pattern for thermal sprayed coatings.

• The surface when visually inspected will be free of oil, grease, dirt, mill scale, rust, corrosion products, oxides, paint or any other foreign matter.

• Prepared surfaces will be handled only with clean gloves, rags or slings. Contact with any oil or grease will result in failure of the coating.

• Blasting will not be so severe as to distort the component being prepared for the thermal spray process.

Abrasive Grit

The effects of grit blasting depend on the type and size of abrasive. Sharp, hard, angular particles provide the best results. Spherical or rounded particles should not be used. All abrasives must be clean, dry, and free of oil, or other contaminants.

The most common types of grit used in thermal spray are aluminum oxide and chilled iron grit. These are the only two materials that are approved under DoD-STD 2138 for the final anchor-tooth blasting prior to the thermal sprayed aluminum coating. Garnet and copper slag are generally used for precleaning prior to the final anchor-tooth preparation. They are used primarily due to their low cost. Since the roughness of the finish depends on the size of the grit, abrasives are furnished in different grades. Smaller sized particles will allow for the preparation of more area per hour. Larger abrasive particles produce rougher finishes. It is recommended that the particle size should be 16 to 40 grit for the final preparation for thermal sprayed aluminum coatings and 16 to 60 grit for strip or precleaning blasting.

Adhesion bond strength normally increases with surface roughness, although the value does not improve above the maximum bond strength once the surface profile of three roils is achieved. The adhesion strength conversely drops off dramatically if the anchor-tooth pattern is below two roils. In accordance with DoD-STD 2138, the acceptance criteria for an anchor-tooth profile is two to three roils. There is some controversy in the industry concerning profiles greater than three roils. Profiles over three roils do not increase the bond strength and theoretically require more processing time and cost to (a) achieve a deeper anchor-tooth than required, and (b) apply the TSA coating to fill the profile completely.

There is, however, no loss of effectiveness of the TSA coating system due to excessive anchor-tooth profiles. In no case should an anchor-tooth profile in excess of three mils be considered detrimental to the TSA coating system. Proper training of operators to achieve the required 2-3 mil profile seems the most practical solution.

Blasting Procedures

Besides the abrasive grit type and size, other process variables of importance are air pressure, blast angle and distance. Air pressures for blasting are between 30 to 100 psi depending on the substrate material, flow, size of abrasive particles and the machine and nozzle type used. Low blasting air pressures should be used for very thin substrates to minimize the likelihood of warping damage to the component. With pressure type blasting equipment, 50-60 psi at the nozzle should be used. These are not the pressures at the blast machine tank, but at the blast nozzle as measured with a needle probe gauge. When syphon blasting (suction blasting), the maximum nozzle pressure should be between 75-80 psi.

The compressed air supply should be adequate to furnish the necessary pressure and volume to sustain the proper blast quality. Refer to Table 5-D. The air should be free from oil, water, and other contaminates. Besides clogging the system, oil or water in the compressed air can adversely affect surface preparation and subsequent bonding. The requirements for air quality are defined in DoD-STD 2138. The air equipment used in the abrasive blasting process shall furnish air that is free of oil and moisture (less than 0.03 parts per million (p/m) oil). The abrasive stream should be directed onto the substrate surface at the same spray angle as the application of the TSA coating, which is normally between 45-90 degrees. Nozzle to substrate distance varies from 4- 12 inches depending upon the size and type of abrasive used, nozzle opening size, and capacity of the blast machine.

Blasting Rates

Blasting rates depend on several factors including the type, size, and loading capacity of the blasting equipment as well as the substrate material. Blast machine nozzles having large diameter orifices will cover more area per hour than nozzles having smaller orifices. However, the size

NOZZLE DIAMETER, IN.		NOZZLE PRESSURE, PSI					
		50	60	70	80	90	100
1/8	cfm air	11.3	13.2	15.1	17.0	18.5	20.3
	lb/hr	67.0	77.0	88.0	101.0	112.0	123.0
	hp required	1.6	2.1	2.5	3.1	3.5	4.2
1/4	cfm air	47.0	54,0	61.0	68.0	74.0	81.0
	lb/hr	268.0	312.0	354.0	408.0	448.0	494.0
	hp required	6.4	8,3	10.2	12.4	14.2	16.8
3/8	cfm air	103.0	126.0	143.0	161.0	173.0	196.0
	lb/hr	668.0	764,0	864.0	960.0	1052.0	1152.0
	hp required	14.8	19.3	23,9	29.3	33.2	40.6
1/2	cfm air	198.0	224,0	252.0	280.0	309.0	338.0
	lb/hr	1160.0	1336.0	1512.0	1680.0	1856.0	2024.0
	hp required	26.7	34.3	42.1	51.0	59,3	70.0

Table 5=D Air, Grit Flows and Power Requirements for Blast Nozzle Sizes of the nozzles to be chosen is limited by the amount of compressed air available. The type and size of the abrasives also influence blasting rates. Generally, the larger the abrasive particle size, the slower the operation. Approximately fifteen pounds of aluminum oxide or twenty-five pounds of chilled iron grit are required per square foot of blasted surface to achieve a properly prepared surface. Approximate blasting rates for various equipment types are shown in Table 5-E.

Grit Recycling

Some abrasives used in shop or production applications may be recycled, cleaned, and screened so that they can be used again. Angular chilled iron grit and aluminum oxide are most commonly used in such operations. When an abrasive is reused, it should be cleaned of dust and resized, with a minimum of 80% conforming to the original size requirements. Contaminated abrasive grains, or those of questionable quality, should not be reused. Failure to remove broken down grit (fines) from the abrasive can be detrimental to the proper bonding of a coating.

GUIDE TO BASE METAL GRITBLASTING RATES FOR NACE* NO. 1, WHITE METAL FINISH		
	RATE, FT²/HR	
EQUIPMENT	MINIMUM	MAXIMUM
Pressure type	20	40
Manual abrasive return room	30	60
Automatic abrasive return room	50	160
Wheel type airless (per wheel)	150	400

Notes:

The minimum rate indicates the output on small work heavily corroded requiring considerable handling.

The maximum rate is for large surfaces in semi-bright or lightly corroded conditions.

*National Association of Corrosion Engineers

Table 5-E Gritblasting Rates

Table 5-F shows frequently used materials evaluated under controlled conditions. There are many variables that can alter this data, but the table gives a relative idea of the number of times different types of material can be recycled through a blast machine.

ABRASIVE LIFE	
MATERIAL	NO. OF RECYCLES
Aluminum oxide	10
Chilled iron	15
Steel	100
Garnet	

Table 5-F Abrasive Recycling

Provisions should be made to incorporate enclosures to capture the spent abrasives used in on-block or on-ship applications. In addition, reclaiming/cleaning devices should be used wherever possible to maximize grit usage.

5.4 TSA Coating Procedures

The thermal sprayed aluminum process is operator technique dependent. only personnel who have been certified to Dod-STD 2138 and have received the proper training should operate the spray gun, and all process instructions must be strictly adhered to. Prior to application of thermal sprayed aluminum, all areas to be coated must be pre-planned in regard to spraying technique. Consideration must be given to the potential for overspray at an improper gun angle or distance, in addition to following all of the required parameters.

Manufacturers of thermal spray equipment have defined procedures to properly set-up the equipment. General procedures are as follows:

• Review parameter records.

• Ensure gun has appropriate hardware.

• Check, clean and connect all hoses and cables to gun.

• Test run system to ensure operation is correct.

Operator Technique

Prior to the production application of TSA coating, the operator is required to demonstrate that the equipment and process techniques are correct by preparing and spraying a test sample (3 inches by 2 inches by 0.050 inch) with 7-15 mils of aluminum. These samples are called Bend Test Coupons and they serve as a pre-production verification that all procedures are being correctly followed. The anchor-tooth profile and coating thickness are recorded daily on a shop traveler or log, showing the items that will be processed by that particular operator. Each sample is bent in accordance with the requirements of DoD-STD 2138. Results are evaluated and recorded. (Refer to Section 6 for acceptance criteria and procedures for the Bend Test.)

In the event the Bend Test fails to meet the acceptance criteria, no processing can commence until the procedure or equipment has been rechecked to determine the cause. Once the deficiency has been corrected a second test panel can be processed and tested. If an operator cannot successfully produce an acceptable Bend Test sample the second time, that operator is required to be retrained and his certification will be revoked until he has been retrained and retested. (Refer to DoD-STD 2138, Para. 5.4.2.1) The operator cannot spray any production components until he has been recertified. Prior to starting TSA coating the operator should mentally pre-plan the spraying techniques he will employ to produce the highest quality application possible.

The operator should look for the following

• An identifying area to start the coating, and be able to come back to that area for the second or third pass to ensure coating coverage.

• Areas which require angle heads or extensions.

l Areas where potential overspray at the wrong distance or spray angle will cause poor adhesion or excessive porosity.

• Areas where the masking may deteriorate due to the heat generated by the coating.

• Sharp edges or corners where the coating may be chipped or damaged if the maximum coating thickness is applied. Ideally, all edges and comers should be rounded or radiused prior to coating. Edges are the weak link in any coating process.

Each operator should develop the skills to determine audibly or visually if a thermal spray gun is operating correctly. This audible skill may take years to perfect but experienced operators can tell if a gun is operating correctly by the pitch the gun emits during processing. The visual skills are more definable; the following list describes the visual characteristics operators should be capable of determining:

• The spray texture of the deposited coating

• Flame or arc color

• Flame definition (sharpness)

Ž Length and condition of the melting tips of the wire

• Water or oils present on the gun or hoses

Operators also should develop the skills to recognize variation from the correct operating parameters and have the ability to correct them. During the coating application, operators should monitor the operating parameters, i.e., pressure/flows or voltage/amperage to detect any variations in the pre-established values. The TSA coating process success is directly proportionate to the skill, knowledge level and dedication of each operator. Most thermal sprayed coating failures occur due to poor operator technique. All application parameters have a tolerance range within which the TSA coating will be acceptable. If the operator keeps the application techniques in this range, he can be assured of the success of the coating.

Coating Characteristics

An inherent characteristic of the coating process is the stress created by the cooling of each particle during spraying. Stress is the combination of thousands of particles shrinking due to the quenching effect of the molten particles impinging on the cool work piece. The coefficient of expansion and contraction of aluminum is substantially greater than other materials. The stress factor can be controlled but if it is not, stresses can exceed the bond strength of sprayed aluminum and the coating can disbond.

The combination of the bond and cohesion strength make up the overall bond of the coating to surfaces. The difference between bond strength and cohesive strength is as follows:

Bond Strength That first layer of particles that are in intimate contact with the prepared base materials. The bond strength should always be greater than the cohesive strength of the coating.

Cohesion Strength The adhesion strength of particles to particles generally defined as cementation.

To keep the stresses at a minimum, the aluminum coating should be applied in multiple passes that will reduce the number of particles impacting any one area. The gun movement relative to the target is the controlling factor. Each operator, through hands-on training, will be able to optimize their technique for applying the coating in multiple passes. The spray pattern or overlap between passes is another technique the operators must perfect. For combustion spraying with a standard air cap, the coating will be applied evenly with overlaps of 5/8" - 3/4." There are fan air caps available for the combustion wire systems to extend the spray pattern to 2" - 3" wide. The fan air caps should only be used for large flat surfaces.

For arc spraying with standard air caps, the coating will be applied evenly with overlaps of approximately 1" - 1½." With fan air caps, the spray pattern can be increased up to 6" and overlaps up to 5" will produce even coatings. The fan air caps should again, only be used for large flat surfaces.

Angle air caps, nozzles and extensions are readily available for the combustion powder and wire systems, with limited availability for the arc spray systems. These angle spray devices permit the coating of areas where a straight spray pattern will not cover adequately. They provide up to a 45 degree spray angle while the gun is pointed straight. The extensions can be very helpful to spray areas where an operator's arm cannot reach. These extensions are manufactured up to a standard length of three feet, although longer lengths can be special ordered. For both combustion wire or powder and electric arc systems, the use of angle heads and extensions reduce the pounds per hour capabilities by approximately 20%.

Correct overlap is an important technique to ensure an even deposit across any given surface. In any hand-spraying application, it is virtually impossible to maintain the required overlap consistently. One technique used to help maintain an even coating deposit is to make the second or "crossing pass" at right angles to the original pass. This technique will maximize the efficiency of the coverage.

When applying TSA coatings over blasted steel surfaces, there is a slight color difference between the TSA and the blasted surface. (The TSA is white and the blasted surface is light gray). Operators can visually see where they have coated. However, making subsequent passes over previously TSA coated surfaces is more difficult since there is no visual way an operator can determine where the subsequent passes have started or stopped. The only way to determine the coverage is by measuring coating thickness. Therefore, it is strongly suggested that only small areas (approximately two square feet) are processed at a time. This will allow an operator to concentrate, and not lose track of where he started or where the next overlap pass will start.

The drawback to processing small areas is the potential for other prepared surfaces awaiting the TSA coating to become contaminated. Also, the time between blasting and coating may be a factor if the allowable time per DoD-STD 2138 has run out and the surfaces are required to be reblasted. In some cases, to eliminate the requirement of reblasting any surface, a flash coat of 1 - 2 roils applied over the prepared surface will prevent unnecessary duplication of processing steps.

Coating Application

A localized preheating procedure to eliminate any surface moisture on the substrate is recommended prior to application of TSA coatings. A preheat temperature of 125-150 degrees Fahrenheit is a good standard practice for thermal spraying. The surface temperature should not exceed 350 degrees Fahrenheit any time during the spraying process. If the steel substrate temperature of the item to be sprayed is less than 10 degrees Fahrenheit above the dew point, no thermal spraying should be conducted.

To ensure proper application, a spray distance of 5" - 8" at a 90 degree angle (perpendicular) is the optimum condition. A spray angle of 45 degrees is the minimum acceptable angle. Beyond a 45 degree angle, the particles will shadow the next particle and prevent it from flattening out. This causes excessive porosity within the coating structure and poor bond strength and cohesion strength.

After anchor-tooth blasting, the TSA coating should be applied to the desired surface as soon as possible. DoD-STD 2138 mandates that in no case should a component be sprayed after six hours has elapsed. The TSA coating operation should be started within four hours after final anchor-tooth blasting, and finished within Six hours. Parts not sprayed within six hours are to be lightly reblasted to remove surface oxidation and then sprayed. If any flash rust, discoloration or contamination occurs on a blasted surface, that part must be reblasted to white metal prior to continuing the spray process.

The optimum coating thickness per pass is 3-4 mils and the optimum total coating thickness is 7 - 10 roils for parts whose operating temperatures are below 175 degrees F. For parts whose operating temperatures are above 175 degrees, the optimum total coating thickness is 10-15 mils. The rate of gun movement and indexing (overlapping) will create the proper **thickness per pass.**

	HIGH TEMP. (TYPE I)	LOW TEMP. (TYPE II)
Optimum	10-15mils	7-10mils
Acceptable	10-20 roils	7-20 roils

Table 5-G Coating Thickness Range

The coating thickness should be measured with a magnetic or electronic thickness gage. Each operator should initiate their own quality control checks prior to any formal inspection. It is recommended that the operators perform preheat temperature and thickness readings during the processing, and that they thoroughly check for coating thickness prior to final inspection.

In any case, if the coating thickness is greater than 20 roils, the coating must be removed by abrasive blasting and then resprayed to the proper thickness. If the coating thickness is less than 7 mils and no surface contamination has taken place, additional coating can be applied to obtain the required thickness. If surface contamination occurs, a light blast should adequately prepare the surface and the part can be resprayed. If the contamination consists of oils or grease, solvent degrease and remove all residue and prepare the area by lightly blasting. The part can then be resprayed.

Sealing the TSA Coating

Thermal spray coatings are inherently porous. Porosity can range from less than 1% to greater than 15%. If the coating is improperly applied, particularly with complex shapes, this porosity may extend from the coating surface to the substrate. Sealers are used as a post treatment to fill such pores. Reasons for sealing sprayed coatings are:

- Prevention or retardation of corrosion at the coating/substrate interface.
- Life extension of aluminum corrosion preventive coatings.

Sealers must be applied to clean, dry, flame-sprayed surfaces. Any oil, grease, or other contamination on the aluminum coating must be removed by washing with a thinner compatible with the sealer.

Use TT-P-28 (environmentally compliant) heat resistant aluminum paint on sprayed surfaces of components operating at temperatures greater than 175 degrees. A 1.5 mil dry film thickness is required per coat for a total of 3 roils. For components whose operating temperature is less than 175 degrees F., use Formula 150 (MIL-P-24441-IC) epoxy at one mil or less for the sealer coating. Dilute the epoxy 50% by volume with an approved epoxy thinner before applying the seal coat. (NOTE: This thinned epoxy sealer will be specified in the revision to DoD STD 2138.) Subsequent full coat paint systems compatible with the diluted Formula 150 are applied over the sealer.

7. PROCESS CONTROL

7.1 Quality Assurance

Quality assurance is the responsibility of all personnel directly or indirectly involved with the thermai sprayed aluminum (TSA) coating process. Each shipyard has their own quality system, and the corporate management objectives regarding quality should be clearly defined. Implementation and control is the responsibility of all members of the particular organization. The effectiveness of the TSA coating process will only be successful if the commitment to quality is maintained.

The operator technique dependency of the TSA coating process demands that each operator is responsible to meet or exceed the acceptance criteria for the coatings. Each step in the processing is a key ingredient to producing a quality product. Any steps not correctly performed within the requirements of DoD-STD 2138 may lead to premature coating failure, thus jeopardizing both an organization's credibility and the validity of the TSA coating process throughout the industry. The only way an organization can continually produce acceptable TSA coatings is to create reliability and repeatability in processing with an effective quality control system.

The first step in establishing a quality control system for the TSA coating process is education. Within the organizational structure of a typical shipyard, interfacing with all trades during new construction or repair is complex enough without having to administer a new process (TSA) or include this process into a repair work scope.

Each trade, Planning Department, Quality Assurance (QA) Department, Safety Department, and Production Department should be familiar with the TSA coating process and how it will affect the components or areas they are responsible for. The quality of a TSA coating after being originally applied, can be jeopardized by mechanical damage from welding, cutting or handling. Each trade should have the knowledge to deal with areas or components that have been TSA coated. Another part of the educational process is the training of operators, inspectors and supervisors to understand how the TSA coating process acceptance criteria is achieved. The careful selection of personnel to accomplish this process requires operators to have the skills, discipline and commitment to produce consistent quality work. Each potential operator should be pre-interviewed and assessed to determine if they have the required skills and necessary characteristics.

Quality Requirements for Production

Many shipyards have designated production personnel to perform a majority of the quality checkpoints. The Quality Assurance Department validates and performs key checkpoints such as, surface preparation and coating thickness. Each shipyard should define the individuals or departments for the following inspection responsibilities.

<u>Receiving Inspection</u>

Ž Initiate control documents (shop traveler, Q A reports).

• Ensure components are disassembled properly.

• Identify areas not to be sprayed such as: dissimilar metals (ferrous only are sprayed), machined surfaces, weld zones, close tolerance fits, key ways and sliding surfaces.

<u>Masking Inspection</u>

• Ensure all areas not to be blasted or sprayed are properly and sufficiently protected using appropriate masking materials.

Anchor Tooth Blast Inspection

- Ensure all masking materials are intact.
- Ensure part has not been damaged by blasting.
- Ensure a clean white metal surface (SSPC-SP5).
- Ensure an anchor tooth profile of 2-3 mils (0.002-0.003 inches). Record results.

Pre-production Bend Test

- Verify and record anchor tooth profile.
- Verify and record coating thickness.
- Prepare, TSA coat, test and evaluate one bend test coupon for each operator per shift.
- Compliance of the bend test samples is required prior to production work.

Temperature and Humidity Requirements

- Measure and record ambient temperature.
- Measure and record relative humidity.
- Determine dew point.
- Measure and record component temperature, ensure component temperature is a minimum of 10 degrees F. above the dew point.

Coating Inspection

- Measure and record coating thickness (5 random areas).
- Ensure total coverage of the TSA coating.
- Visually inspect for any blisters, chips, cracks or surface contamination of the coating.
- Visually inspect using a 10X magnifying glass for nodules not exceeding the acceptance criteria.

Final Inspection

- Ensure parts have received proper sealer coat and paint system.
- Ensure parts are properly packaged for transport.
- Review records package for traceability documentation.

Facility Quality Assurance Requirements

Calibration System: Maintain all gases, flow/amp/voltage meters, measuring tools and testing equipment in current calibration in accordance with Mil-Std 45662A. These devices should have up to date calibration stickers showing the next and last calibration dates.

Air Quality check Periodically, (yearly at a minimum) have the quality of the air used to support the TSA process checked by a laboratory for compliance with DoD-STD 2138. This certification could be checked more frequently if there is doubt the dryer system is functioning adequately.

Operator Certification Maintenance: Establish a recall system to control each operator's TSA certification by reviewing their performance and coating production records. This will ensure they have met the minimum requirements of maintaining certifications.

Grit Qualitv Testi Set up a daily procedure to verify and record the condition of the grit used for anchor-tooth profile blasting. Refer to the sample in Appendix G.

Records

DoD-STD 2138 defines the record keeping requirements as follows:

Records of facility approval, application procedure, operator certification test results, and production records, shall be maintained by each performing activity, contractor or subcontractor. These records shall be available to the contracting agency for review and audit. The performing activity shall maintain these records for six months after completion of the contract work

Copies of the records shall be made available to the contracting agency upon request.

Operator Certification Test Sample Requirements (DoD-STD 2138)

To comply with the requirements for certification or recertification, each operator must prepare, TSA coat and test the following samples

Four Bend Test Coupons
Five Bond (Tensile) Test Samples
Two Shape Test (1 "T" and 1 Round)

Each of the above samples will be examined in accordance with the visual acceptance criteria. The bend test and the bond (tensile) test samples are required to meet the acceptance criteria for their respective tests.

Test Sample Sizes and Materials

Bend Test: 0.050" thick x 2" x 3" minimum, flat panel (mild steel) 4130 or equivalent

Tensile Test Coupons: 1" Diameter x 2" long material (mild steel) 4130 or equivalent

Shape Test Samples, "T": 1/2" thick, base 6" long x 3" wide, leg 3" high x 3" wide

Shape Test Sample, Round 1/4" wall thickness, 2" diameter x 6" long material (mild steel) 4130 or equivalent

TSA Coating Test Procedures and Acceptance Criteria

Tensile Test: Tensile tests are required at the time of operator qualification or requalification. Tensile test specimens shall be machined and prepared according to the requirements of ASTM C 633 and DoD-STD 2138. The acceptance criteria of the coating is a minimum tensile strength of 1500 psi and an average tensile strength of 2000 psi. Tensile tests shall be accomplished per ASTM C *633*. Fig. 7.1 shows typical failure modes for tensile test specimens.

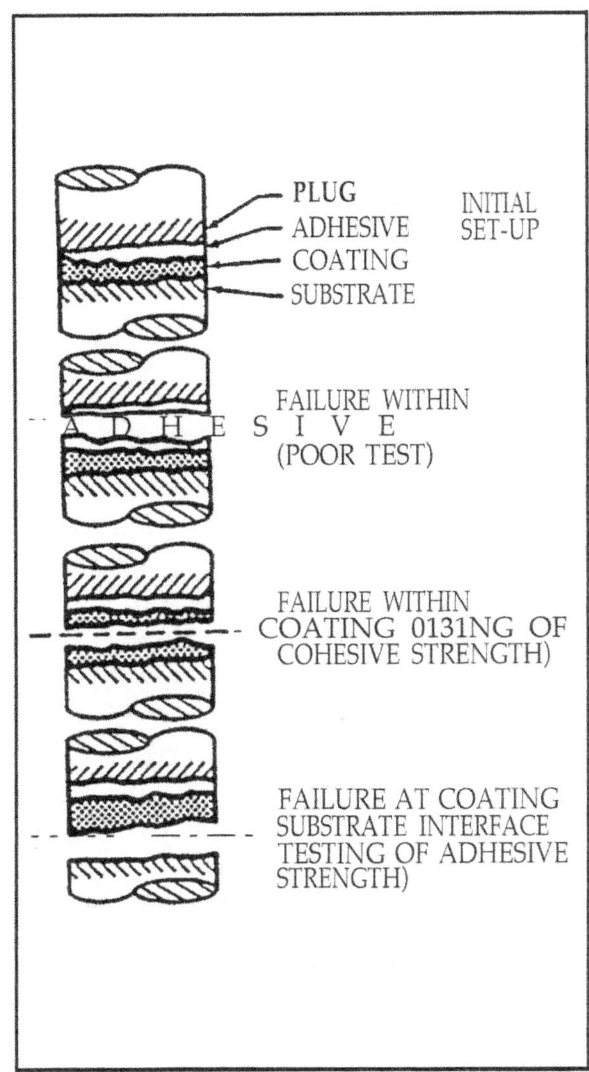

Fig. 7.1 Modes of Coating Failure

Bend Tests After preparing and applying the coating on one side of the bend test sample, insert sample with the coating on the outside into a bend test fixture and bend the sample 180 degrees around a 1/2" diameter mandrel. Visually examine the coating for disbanding or delamination caused by the bending. Small cracks and "alligatoring" of the coating in the vicinity of the bend are acceptable. Refer to Fig. 7.2 or Appendix H for acceptance criteria.

Visual Inspection Requirements: Each sample item should be examined visually at

ACCEPTABLE

MINIMALLY ACCEPTABLE

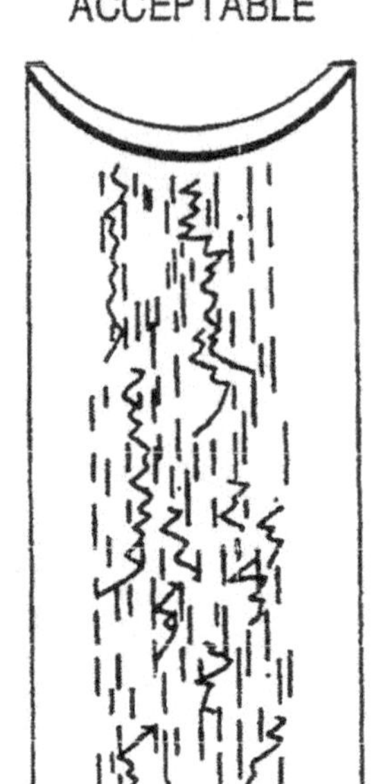

NOT ACCEPTABLE

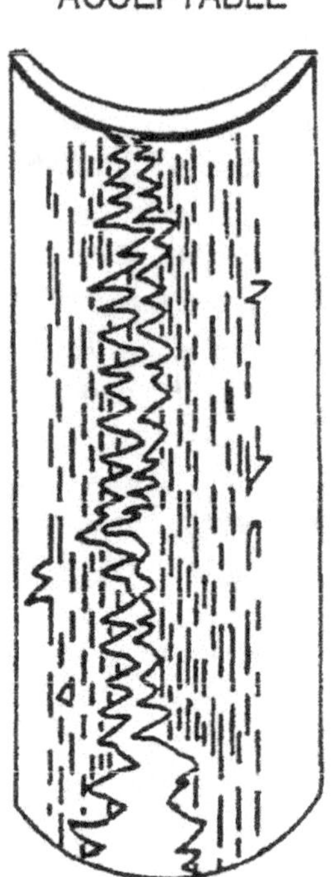

Fig. 7.2 Bend Test Samples

a magnification of 10X. The coatings should have a smooth, uniform appearance. The coating system should not contain any cracks, pin holes, or chips that expose the metal substrate. Surface defects of the TSA coating should be limited to small nodules not to exceed 0.045 inches in diameter, and should not exceed 0.025 inches in height above the surrounding sprayed surfaces. The coating should not contain any of the following:

- Blisters
- Cracks
- Chips or loosely-adhering particles
- Oil or other internal contaminants
- Pits exposing the undercoat or substrate

In the event of a suspected non-compliant coating application, a knife peel test or adhesion test can be used to determine if the coated area will meet the minimum acceptance requirements. The adhesion test is applicable only to interior wet spaces as defined in DoD-STD-2138 page 1 Para. 4.4.5, Category III. (See Appendix H.)

7.2 Coating Repair

Components being transported from the spraying area to the installation area must be handled carefully to avoid damaging the surface. Large items should be lashed to a pallet for shipment. Smaller items should be placed in boxes with the items wrapped or separated.

Extreme care must be exercised when installing or assembling the component. If small nicks or gouges are found, the item should be cleaned, resealed and top coat painted as soon as possible. Small nicks and gouges can be touched-up without harming the life of the coating. However, large dents in the aluminum coating will require removal of the damaged area and reapplication of the TSA coating and sealer.

Although the new revision of DoD-STD 2138 has not been officially released as of this writing, proposed new specification requirements to repair TSA coating systems are defined in Table 7-A.

SUBSTRATE NOT EXPOSED		SUBSTRATE EXPOSED	
SMALL AREA (<100 IN²)	**LARGE AREA (>100 IN²)**	**ONLY PAINT TOUCH-UP REQUIRED**	**TSA COATING REPAIR REQUIRED**
• Solvent clean as required. • Use a 1" flexible-blade paint scraper, and remove loose paint around worn or damaged area to the boundary or the well bonded paint. Take care not to gouge or further damage the TSA coating. • Use a stiff hand-held non-ferrous bristle brush and vigorously brush away loose debris. • Feather a 2-3" collar into the undamaged area. • Lightly abrade the feathered paint area around the exposed TSA coating with sand paper to provide a mechanical bonding surface for the paint primer and sealer.	• Solvent clean as required. • Abrasive brush blast away loose paint using aluminum-oxide grit over the exposed TSA coating area. • Feather a 2-3" collar into the well bonded paint area. • Minimize cosmetic difference between new and old paint by brush blasting or using sand paper and repainting as area bordered by a weld bead or a structural item.	• Solvent clean as required. • Using a paint scraper, push the blade underneath the TSA coating to lift off all loosely bonded TSA coating until reaching a well-bonded area. • Use portable disc sander with 80 mesh sandpaper and clean steel substrate to clean metal feathering 2-3" into the undamaged coated area.	• Solvent clean as required. • Abrasive blast area to be repaired with 16-30 mesh aluminum oxide to white metal to give a 2-3 mil anchor tooth. • Feather 2-3" into the good coating area. • Apply TSA coating as specified.

Table 7-A Proposed TSA Repair Procedures From Draft 2138(A)

In cases when areas have purposely been left unsprayed, such as weld zones, touch-up on these areas on-board ship may have to be accomplished. The ideal method for this procedure is as follows:

• When originally applying the TSA coating, mask the anticipated weld zone.

• Apply the TSA coating up to the masked area.

• Remask approximately 1 inch above the boarder of the TSA coating leaving a boarder of aluminum.

• Seal the TSA coating to the masked boarder.

• Install the component on board ship or on block.

• Remove the masking, and blast and TSA coat the unsprayed areas, using extreme caution when blasting on to the existing TSA coating.

l Reseal the newly applied TSA coating.

The reason for this procedure is to avoid sealing any area where additional TSA coatings are to be applied and to prevent contamination of the tie-in area where the new coating meets the old.

The reblasting of any TSA coating for the purpose of recoating or tying-in areas adjacent to those that have been previously coated requires a delicate technique. Blasting existing TSA coatings is not difficult, but it needs to be handled differently than when blasting uncoated surfaces. The TSA coating will tend to work harden under the high pressure and impact of normal blasting techniques. Reduced air pressures and faster, longer distance passes are necessary to avoid damage to the coating. Practice during the hands-on portion of the training will greatly enhance this technique.

Complete removal of an existing, properly applied TSA coating by grit blasting is a very difficult, time consuming process. It is best to use a smaller mesh size abrasive (60 - 80 grit) at an angle of approximately 15-20 degrees to peel the coating from the surface.

There are chemicals that can readily remove the aluminum if the component is small enough to be immersed within a tank. Prior to using any caustic chemicals for coating removal, check with the shipyard's metallurgist to ensure the chemical will not damage the base material. There is also a possibility of the chemicals being entrapped within welded or porous areas of the component, and when a new TSA coating is applied, the chemical could react with the coating and start to deteriorate the coating from the inside-out. Residual chemicals could be removed by high pressure water blasting, although high risk areas should not be stripped using chemicals without ensuring complete removal of the chemicals.

In cases where large areas of aluminum coating require removal, high pressure water jet blasting is a cost effective removal technique. An in-depth study was performed by Puget Sound Naval Shipyard, and can be obtained from NAVSEA Code 07011. (Also see Reference 1.)

11. THE FUTURE OF THERMAL SPRAY IN SHIPBUILDING

Thermal sprayed aluminum (TSA) coatings have been proven to be a cost-effective means of providing long-term corrosion protection to many shipboard components. For this reason, the U.S. Navy is expected to continue specifying the application of these coatings for future new building programs, as well as repair and overhaul contracts. The Navy is also investigating new advancements in thermal sprayed coating technology as described below.

The Navy has recently authorized the use of TSA coatings 3-4 roils thick (in place of the current 7-10mils requirement) followed by a top coating of TGIC (Triglycidal Isocyanurate) Polyester powder coating, electrostatically applied and oven-cured, for topside component applications. The new revision to DoD-STD 2138 will cover this application. The procedure provides a coating system that is superior to the existing five-coat topside paint system. The use of TGIC Polyester as a top coating eliminates the VOC environmental issues of the five coat conventional paint system and reduces the processing time of the barrier coat system from five days to two hours.

Powder coatings of TGIC Polyester alone are also being specified for many other shipboard components. For shipyards wishing to investigate the potential of setting-up powder coating equipment, there are three pieces of equipment required

- A powder coating booth and reclaiming system.
- An electrostatic powder coating application system.
- An oven capable of handling the largest size component to be processed with a maximum temperature range of 350-400 degrees F.

There are new thermal spray coating systems on the horizon. These systems are designed as duplex coatings, consisting of a TSA coating, top coated with a thermal sprayed thermal-plastic or TGIC Polyester. This system would allow large surfaces or on-board items to be coated with a superior topside corrosion control coating system, virtually eliminating the requirement for a five coat paint system. Bilges, tanks, and wetspaces, as well as interior and topside areas are being considered for potential application of these advanced coating systems.

In commercial applications, thermal spray coatings are widely accepted for corrosion control. These coatings provide long-term corrosion protection from many environments. The life cycle analysis and evaluations indicates thermal spray coatings to be extremely cost efficient.

The major oil companies such as Shell Oil and Conoco have developed comprehensive application specifications. They are convinced that thermal spray coating can out-perform conventional corrosion control methods in both submersion and splash zone environments for many of their offshore platform's high corrosion prone areas. The potential scope of this coating work is large, including new and existing platforms. Most United States shipyards can provide the oil companies with proposals to build structural steel platfoms, and TSA coatings will be part of the corrosion control system used to protect the structure. Shipyards can benefit by having the capability to provide these coatings to support the offshore industry.

www.ingramcontent.com/pod-product-compliance
Lightning Source LLC
LaVergne TN
LVHW080320110826
845155LV00026B/164

* 9 7 8 1 9 3 4 9 3 9 1 1 6 *